CULTURE AND COSMOS
http://www.CultureAndCosmos.org

Culture and Cosmos is published twice a year, in northern spring/summer and autumn/winter, in association with the Sophia Centre for the Study of Cosmology in Culture, University of Wales Trinity Saint David.
Contributions and editorial correspondence should be addressed to:
Editors@cultureandcosmos.org

Editor: Dr. Nicholas Campion, the Editor of *Culture and Cosmos*, School of Archaeology, History and Anthropology, University of Wales Trinity Saint David, Ceredigion, Wales, SA48 7ED, UK.
E Mail **n.campion@tsd.ac.uk**

Deputy Editor: Dr. Jennifer Zahrt
Assistant Editor: Dr. Fabio Silva
Editorial Board: Dr. Silke Ackermann, Professor Anthony F. Aveni, Dr. Giuseppe Bezza, Dr. David Brown, Professor Charles Burnett, Dr. Hilary M. Carey, Dr. John Carlson, Dr Patrick Curry Professor Robert Ellwood, Dr. Germana Ernst, Dr. Ann Geneva, Professor Joscelyn Godwin, Dr. Dorian Greenbaum, Dr. Jacques Halbronn, Robert Hand, Dr Jarita Holbrook, Professor Michael Hunter, Professor Ronald Hutton, Dr Peter Kingsley, Dr. Edwin C. Krupp, Dr. J. Lee Lehman, Dr. Lester Ness, Professor P. M. Rattansi, Professor James Santucci, Robert Schmidt, Dr. Lorenzo Smerillo, Professor Richard Tarnas, Dr. Graeme Tobyn, Dr. David Ulansey, Robin Waterfield, Dr. Charles Webster, Dr. Graziella Federici Vescovini, Dr. Angela Voss, Dr. Paola Zambelli, Robert Zoller.
Technical assistance: Frances Clynes
Copy-editing: Ian Tonothy, Marcia Butchart.

Contributors Guidelines: Please see http://cultureandcosmos.org/submissions.html
Copying: Apart from fair dealing for the purposes of research or private study, or criticism or review, as permitted under the Copyright, Designs and Patents Act 1988, no part of this publication may be reproduced, stored or transmitted in any form or by means without the prior permission of the Publisher.
Front cover: Sunset over Faial island (Azores, Portugal) eleven days after the summer solstice, photo courtesy of Fernando Pimenta (2013) (see 'Land, Sea and Skyscape: Two Case Studies of Man-made Structures in the Azores Islands', pp. 107–32, this issue).

Published by Culture and Cosmos, PO Box 1071, Bristol BS99 1HE, UK.
© **Culture and Cosmos 2015**
Printed by Lightning Source

CULTURE AND COSMOS
www.CultureAndCosmos.org

Editor Nick Campion
Vol. 17 No. 2 Autumn/Winter 2013   ISSN 1368-6534

Published in Association with
The Sophia Centre for the Study of Cosmology in Culture,
University of Wales Trinity Saint David
http://www.uwtsd.ac.uk/sophia/

Editorial

Landscape – Seascape – Skyscape

This issue of *Culture and Cosmos* stems out of the 'Land, Sea and Sky: a "3-scape" approach to Archaeology' session of the 2013 meeting of the Theoretical Archaeology Group (TAG), at the University of Bournemouth. The session, which was organised by myself, was intended 'to extend the discussion on Landscape and Seascape Archaeology, by considering their interaction with a third "-scape": the skyscape'. This was the second such session at a TAG meeting, the first of which occurred the previous year and resulted in the volume *Skyscapes: The Role and Importance of the Sky in Archaeology*.[1]

These sessions at TAG have been instrumental in bridging the gap between the fields of cultural astronomy and archaeology, bringing cutting edge archaeoastronomical work to the attention of archaeologists and vice-versa. In this, they are part and parcel of a larger movement in British academia, with the University of Wales Trinity Saint David at its forefront. The School of Archaeology, History and Anthropology has recently put forward a five-year research plan based on the trinity of landscape, seascape and skyscape. The MA in Cultural Astronomy and Astrology has also incorporated this notion and retitled and revamped the Archaeoastronomy module, which from 2015 onwards will be known as 'Skyscapes, Cosmology and Archaeology'.

---

1 Fabio Silva and Nicholas Campion (eds), *Skyscapes: The Role and Importance of the Sky in Archaeology* (Oxford: Oxbow Books, 2015).

This issue opens with a theoretical discussion of how the trinity of land, sea and sky conspire to define place. Daniel Brown explores this and highlights its importance for the interpretation of archaeological sites with celestial alignments. He argues that one needs to explore archaeological structures not merely with recourse to statistical methods and plan views, as often happens, but to actually witness its context—the landscape, seascape and skyscape—that was 'witness and motivator to the erection of both monument and alignment'.

This is followed on by Pamela Armstrong who, in her contribution to this issue, asked the question of whether the material record of the Cotswold-Severn region displays evidence for a shift from lunar to solar symbolism at the Neolithic transition. Armstrong looks at a number of fully excavated structures that were built during this key prehistoric transition in search of the celestial alignments that are known to have featured in later monuments, such as Stonehenge and Avebury. Her findings reveal that, unlike the often made, overly simplistic assumptions of lunar or solar centric cosmologies, a 'more varied appreciation of the sky existed in those earliest of times'.

Looking further west into Pembrokeshire in southern Wales, Olwyn Pritchard identifies a pattern in the distribution of prehistoric monuments and settlements that suggests the possibility of a north-south trans-peninsular route linking Carmarthen Bay, in the south, to Cardigan Bay, in the north. This route was marked, at its southern end, by the King's Quoit dolmen that the author had previously suggested was oriented northwards towards the lowest culmination of Deneb and Vega. The route then passes through several other monuments, including the Banc Du causewayed enclosure and associated monumental complex. Prehistoric Welsh sea, land and skies were thus linked in a unique way through what was surely a route of symbolic, if not ritual, significance.

The remaining two papers focus on human-made structures built on island settings. Firstly, Tore Lomsdalen discusses the islandscapes of the megalithic temples of the Maltese archipelago. These islands were only inhabited in late prehistory, presumabily via Sicily, which implies a working knowledge of celestial navigation since the Maltese islands are too far out to be visible. A thousand years after arriving, these colonists developed one of the richest and most unique prehistoric cultures in the whole of Europe, with their vast megalithic temples and associated iconography. Lomsdalen discusses the role played by the sky, in association with the land and the sea, for this mysterious culture.

Lastly, Fernando Pimenta and collaborators present preliminary work carried out among some structures in the Azores islands. These artificial structures have long been ignored by historians and archaeologists alike but have recently been featured in the debate surrounding a possible human presence in the archipelago predating the arrival of the Portuguese in 1432. Pimenta looks at the location and orientation of the pyramidal *maroiços* of Pico island as well as of three peculiar caves of Terceira island, with respect to the surrounding landscape and skyscape. The findings suggest a symbolic element to these structures that goes beyond the purely functional previous interpretations, and justifying the need for further research and excavation.

Also featured in this issue are two book reviews, contributed by two other speakers at the TAG session. Liz Henty reviews Richard Bradley's *The Idea of Order: The Circular Archetype in Prehistoric Europe*,[2] whereas Lionel Sims looks at Mike Parker Pearson's *Stonehenge: Exploring The Greatest Stone Age Mystery*.[3] These are two works from distinguished archaeologists, authorities in their field, that cross-cut and correlate to each other in their ultimate goal of trying to understand why prehistoric Britons and Europeans went to the effort of building megalithic monuments.

The contributions to this issue demonstrate how the study of the celestial environment—the skyscape—complements the more traditional archaeological approaches to landscapes and seascapes, and how approaching this trinity holistically sheds further light into past societies, their beliefs and practices.

Fabio Silva

School of Archaeology, History and Anthropology,
University of Wales Trinity Saint David.

---

2 Richard Bradley, *The Idea of Order: The Circular Archetype in Prehistoric Europe*. (Oxford: Oxford University Press, 2012).

3 Mike Parker Pearson, *Stonehenge: Exploring The Greatest Stone Age Mystery*. (London and New York: Simon & Schuster, 2012).

The Experience of Watching: Place Defined by the Trinity of Land-, Sea-, and Skyscape

Daniel Brown

Abstract: The interpretation of astronomically orientated ancient sites has frequently led to the conclusion that an astronomy cast needs to exist; a group that is thought to be highly educated as apparent within discussions and comments with researchers even if never published. This group within an ancient society would have a deeper knowledge of the movements of the Sun, Moon and stars derived through observations and analysis similar to the modern day scientific methodology and not accessible to the un-initiated. However, cultural astronomy has started to overcome this mistake by describing such sites not as observatories but as places where a certain phenomenon can be experienced and watched.

This paper will discuss this act of watching and how it is closely linked to a definition of place by introducing the notion of the dialectical image. The triplet of land-, sea-, and skyscape offer common themes and characteristics that allow the watcher to critically negotiate the surroundings and experience the place by dwelling therein. All three form a trinity and are actually part of one skyscape that invokes feelings, illustrates tensions and asks for action. Experiencing this unity is essential to watching. At this stage there is no deeper astronomical knowledge required that is only accessible to the initiated few. Cosmic cycles that manifest themselves through the motions of the Sun, Moon, and stars as well as seasonal and tidal rhythms become obvious to everyone. Skyscapes at astronomically orientated sites capture this meaning. When stepping back from observing and engage in watching, the meaning can be recaptured in the trinity of land-, sea-, and skyscape.

Introduction

The following article is an analysis of land-, sea-, and skyscape as seen from the author's perspective and to some extent discussed by Silva.[1] To

---

1 Fabio Silva, 'The role and importance of the sky in archaeology: an

enable the reader to critically engage with this article it is helpful to outline the background of the author and the motivation behind his work.

The author is a trained astronomer that has carried out several astronomy-teaching activities in rural settings and embraced the surroundings as part of the activities delivered. Such additional activities initially enriched the strictly astronomy related teaching, through exploring the surrounding landscape with its geological and archaeological features or visiting unique trades being carried out at bell foundries or by basket makers. In his more recent work, he is fully integrating the outdoor surroundings and utilising it as a classroom in which astronomy, mathematics, and physics can be experienced. The topic of light pollution lies close to the author's heart and his approach to light pollution education integrates the cultural heritage and the place experience. Within this context and his work on the *Astronomy in the Park* project the views and arguments presented here have been developed.[2]

The movement away from the narrow focus of astronomy taking place only within the observatory to astronomy experienced in a real place within a landscape has been the first step to understanding place and possibly offering a solution to some typical dilemmas faced in modern archaeoastronomy.

In the past decades archaeoastronomy has become more acceptable to many, as Ruggles points out, by placing its finding on a sound and solid basis of statistical analysis.[3] The social relevance of these results have only slowly been integrated in recent years. Even authors such as Harding coining the term skyscape for the first time in an archaeoastronomy context still default to statistical methods.[4] Orientations and alignments surveyed at ancient monuments are gathered together into representative groups.

---

introduction' in Fabio Silva and Nicholas Campion (eds), *Skyscapes: The Role and Importance of the Sky in Archaeology*. (Oxford: Oxbow Books, 2015), pp. 1-8.

2 Daniel Brown, 'How can Higher Education Support Education for Sustainable Development? What can Critical Place-Based Learning Offer?' (MA Dissertation, University of Nottingham, 2013).

3 Clive LN Ruggles, *Astronomy in prehistoric Britain and Ireland*, (Yale University Press, 1999), p. 76.

4 Jan Harding, Ben Johnson, and Glyn Goodrick, 'Neolithic Cosmology and the Monument Complex of Thornborough, North Yorkshire', *Archaeoastronomy*, Vol. 20 (2006): p. 50.

Statistical analysis within these groups allows building up an argument towards intentionality of a measured alignment, and therefore ruling out pure chance alignments. However, such a method can only work if one chooses the correct examples in terms of a typology and regional cohesion i.e., grouping and if there are sufficient examples present. In many cases there might only be one single but unique monument present, for example Stonehenge. But even when working with a large enough group of monuments that have all been built by the same cultural group, the result is a reliable alignment only. Alignment will be either an azimuth or using the local horizon a declination, but it will not tell us what the meaning and motivation was behind establishing this alignment. An alignment might point towards the rising or setting point of the Sun on the horizon at some time during the year. This may also coincide with a Moon rising or setting as well as some of the bright stars. But the alignment alone is not enough to indicate which one of these was the intended target. The situation is made even worse by realising that high precision data analysis is being used giving alignments with well-defined error bars and applying it to buildings constructed by societies that did not either have the technology to achieve high precision or possibly not value precision that much.[5] This could mean that an alignment might only roughly point out an intended setting of a star, since the builder of the monument would know what to anticipate. All these are challenges derive from an inability to extract the meaning behind an alignment using pure statistical analysis and alignment hunting.

Here, it is proposed that exploring the place experience at the site of a detected alignment can help to understand the meaning of it. It will move away from the narrow and scientific term of observation and use the more holistic term of watching. Watching the skyscape along an alignment allows an engagement with a place. This is possible through understanding the landscape surrounding the viewer in a phenomenological way illustrated by Tilley and in a next step as a dialectical landscape that draws you in and communicates with you.[6] To show this an elaboration is required upon what a place actually is and how time is an inherent part of it. This is followed by the description of the dialectical image upon which

---

5 Fabio Silva, 'A Tomb with a View: New Methods for Bridging the Gap between Land and Sky in Megalithic Archaeology', *Advances in Archaeological Practice: A Journal of the Society for American Archaeology* Vol. 2, no. 1 (2014): p. 27.

6 Christopher Tilley, *A Phenomenology of Landscape: Places, Paths, and Monuments* (Oxford: Berg, 1994).

8 The Experience of Watching

the idea of a dialectical landscape has been modelled. The view into the Derwent valley at Gardom's Edge in the Peak District, where the author has worked, will help to illustrate the concept of a dialectical landscape that will then allow formulating how to fully engage and dwell in a place and landscape. Finally, the focus is drawn upon the seascape used as an introduction to becoming one with the place and realising that land-, sea-, and skyscape as well as the viewer are all a unity.

The Place Experience
Since place and its experience is essential in understanding a landscape and becoming more aware of the meaning of a possible alignment, this is the point where the term place needs to be described and the author's view on what defines a place needs to be outlined.

Place is not identical to a location on a map pointing out where one for example lives. There is a big difference between: A Bed and Breakfast accommodation frequented while working away and the home lived in since birth. Both are places where one resides at and should be similar, however the first usually has no personal connection to oneself. It will only be used for the sole purpose of sleeping and having breakfast and it would never be described as a place at which one lived but rather where one stayed. The home is a completely different experience. One lives at home. Every part of it is imbued with memories, feelings, and meanings. One has created part of it and is aware of its history and location within a wider community of neighbours and streets.

The description of place throughout Tuan's book *Space and Place* implies it being filled with meaning and is seen by Heidegger as a locality at which one dwells.[7] As argued by Heidegger, dwelling is a process, at which the surroundings are explored and engaged with, allowing a reflection upon one's existence. When looking upon one's home one can see into one's past and try to make sense of what it means to exist or to be. This might even extend to exploring the past of the house and plot upon which it was built. It might appear as described in Hussey's translation of Aristotle's work on physics III and IV that a place has boundaries similar to a box in which all these meanings and feelings might be collected, but it

---

7 Yi-Fu Tuan, *Space and Place: The Perspective of Experience* (University of Minnesota Press, 1977); Martin, Heidegger, *Being and Time,* (Harper Perennial Modern Classics, 2008): pp. 78–90.

is a far more fluid concept.⁸ Ingold points out that our paths through our lives and the landscape over time intersect and loop around, causing encounters and revisiting past memories. ⁹ These real or imaginary journeys will tangle and create localities without defining a boundary. Such a view picks up on some of the previously mentioned interconnections a place might have with its surrounding, other individuals and history. It also leads on to the important fact that place, though seemingly a purely space related concept, has to contain and is defined by time. Without time, a path could not be followed and memories could not be gathered. Only time helps us to inscribe rhythms in our wanderings and cycles in our lives while carrying out our everyday tasks.¹⁰ Bergson argues, that without time there cannot be emotion and feelings.¹¹ But this time is not the one described by atomic clocks but rather what Sorokin and Merton describe as social time loaded with emotions and memories.¹² The process of drawing one's path is made up of many tasks, described by Ingold in his idea of taskscape, and will evoke feelings and interact with a locality so that both together develop a meaning and a place.¹³

Therefore, it is essential to become aware of time so that one can dwell in a Heideggerian sense thereby realising that both space and time come together to create a place.¹⁴ The time experience can be gathered at a place

---

8 Aristotle, *Physics, Books III and IV,* trans. Edward Hussey, (Oxford University Press, USA, 1983), Phy IV, pp. 212a20–21.

9 Tim Ingold, *Lines: A Brief History* (London: Routledge, 2007), p. 100.

10 Henri Lefebvre, *Rhythmanalysis: Space, Time and Everyday life* (London: Continuum, 2004), p. 8.

11 Henri Bergson, *Time and Free Will: An Essay on the Immediate Data of Consciousness* (New York : Cosimo Classics, 2008), p. 231.

12 Pitirim A. Sorokin and Robert K. Merton, 'Social time: a methodological and functional analysis', *American Journal of Sociology* (1937): p. 622.

13 Tim Ingold, 'The temporality of the landscape', *World Archaeology*, Vol. 25, no. 2 (1993): p. 157.

14 Daniel Brown, 'Skyscapes: Present and Past—From Sustainability to Interpreting Ancient Remains', in Fabio Silva and Nicholas Campion (eds), *Skyscapes: The Role and Importance of the Sky in Archaeology*. (Oxford: Oxbow Books, 2015), pp. 32–41.

by revisiting it over many seasons or over an entire day noticing subtle changes and rhythms as one becomes more aware of the place. Such an engagement would not be focused upon a singular event or single object but has to be open to the entire setting and context of the surroundings at different times. It would be inappropriate to describe this activity or indeed task as observation but it is more likened to watching. The Oxford Dictionary indicative of a common view links observing closely to a scientific approach where something is detected and followed closely.[15] It also contains strong connotations of anticipation, an emotional response heightening awareness. Watching however is defined as looking or looking out for something over a period of time and not focussing necessarily at a specific subject of interest.[16] Only watching includes explicitly the passage of time in its definition. The linking of time to watching becomes even more apparent when noting that its noun is the description of our everyday timepiece—a watch—that helps us to follow the passage of time. And here lies possibly the core of some of archaeoastronomy's recent challenges. Such an experience cannot be accessed through a one-off visit to observe an alignment of only this very specific aspect of a monument. It is this typical modern scientific approach alien to ancient societies that blinds us from discovering a meaning of an alignment and is far from the holistic task of watching.

A Dialectical Landscape
Places containing monuments and their possible alignments are embedded within a much larger network of experiences, emotions and their surroundings. These surroundings are what is termed as landscapes and might be, as in the case of place, associated too closely to the concept of space alone. Here the theoretical concept of a dialectical landscape will be developed, followed by learning how to experience and communicate with a landscape.

---

15 Oxford Dictionary at http://www.oxforddictionaries.com/definition/english/observe?q=observe, [Accessed 28 March 2014].

16 Oxford Dictionary at http://www.oxforddictionaries.com/definition/english/watch?q=watch, [Accessed 28 March 2014].

The Theoretical Outline

The typical places analysed in archaeoastronomy are usually located within a landscape. The following will give a brief overview of the author's understanding of landscape in the art historian context. It is in no way complete and represents the author's discourse with landscape paintings.

The word landscape has its roots in the Anglo-Saxon equivalent of the German word *Landschaft*, denoting a small patch of a larger feudal estate.[17] The term was not used anymore until it became frequented in the arts when referring to depicting a scenery on land.[18] During the antique naturalistic scenes were usually a backdrop to the main topic of the painting. In the Renaissance they became more elaborate and important in the overall message conveyed by the painter, for example in the works by Leonardo Da Vinci.[19] However, the land was never the motive for paintings until towards the seventeenth century Dutch painters made this art form popular. In the following century it spread to other European countries and watercolour paintings of landscapes became an English speciality. In the nineteenth century as romanticism peaked and the industrial revolution was gathering pace, landscape paintings were already popular outside of Europe and were becoming the dominant art. They offered an opportunity to express tensions and hopes arising in artists such as Turner, Cole, and Friedrich. Already the recognition of land as the main motive by Dutch painters as stated by Clark was born from the political religious background of a protestant middle class.[20] A deeper analysis of landscape painting is further motivated for example by the attempt to capture the sublime in giant landscapes created by the Hudson River School and the influence of nationalism on landscape painting trying to describe their national character. Ruskin is one author that has found renewed interest in recent times through the work of Clark, especially in

---

17 Barbara Bender, 'Place and Landscape', in Chris, Tilley et al. (eds.), *Handbook of Material Culture* (Sage, 2006): p. 307.

18 Anon., 'Brief History of the Landscape Genre', J. Paul Getty Museum at http://www.getty.edu/education/teachers/classroom_resources/curricula/landscapes/background1.html [accessed: 13 May 2014].

19 Anon., 'A Brief History of Landscape Painting: Holland Berkley and Igor Medvedev', Park West Gallery at http://www.parkwestgallery.com/archives/14848 [accessed: 13 May 2014].

20 Kenneth Clark, *Landscape into Art* (Harmondsworth, 1956), p. 43.

the context of the interaction of humans and nature.[21] The reading of landscapes used by Ruskin and described in Cosgrove was furthered by others such as Meining.[22] Furthermore, Berger and Williams underline that the analysis has to go beyond a philosophical discussion and must include society and the economy.[23]

One approach to analyse landscapes outlined in more detail here was proposed by Cosgrove and Daniels.[24] Through iconography and a structure of symbols related to power they try to understand the different layers of meaning. They furthermore extend these ideas to analyse landscape gardens, installations, and landscapes in general. They see a landscape as a stage in which power struggles are visible. They illustrate tensions and conflicts between their usage and conservation. Given their starting point of landscape painting it is understandable that they accesses landscapes through physical localities alone placed into a pre-existing setting. But it will become apparent that landscape is far more than just a collection of physical symbols.

Dwelling and discovering one's being requires a dialectic discourse to take place between what one is and is not (A or not A). When viewing a landscape one sees these two extremes ever present—for example in conserving moorland or mining its resources.[25] This dialectic conflict has as an outcome a synthesis apparent in the current landscape; it might be different to our own view of how it should be; at this point one realises that when engaging with these symbols of power one is drawn in and begins to develop a personal conclusion. Initially the process of dialectic was

---

21 Ibid.

22 Denis E. Cosgrove, 'John Ruskin and the geographical imagination', *Geographical Review* (1979): pp. 43–62; D. W. Meining, 'Reading the landscape: An Appreciation of W. G. Hoskins and J. B. Jackson', in D. W. Maining (ed.), *The Interpretation of Ordinary Landscapes* (Oxford, 1979): pp. 195–244.

23 John Berger, *Ways of Seeing* (Penguin UK, 2008); Williams, Raymond, 'The country and the city', Vol. 423, (Oxford University Press, 1975): p. 120.

24 Denis Cosgrove, and Stephen Daniels, *The Iconography of Landscape: Essays on the Symbolic Representation, Design and Use of Past Environments*, (Cambridge University Press, 1989), p. 5.

25 D. A. Gruenewald, 'The Best of Both Worlds: A Critical Pedagogy of Place', *Educational Researcher*, Vol. 32, no. 4, (2003): p. 4.

defined as the compromise between the two only solutions A and not A. The outcome would be a single answer that could then be used for another dialectic discourse that might come closer to the truth of this problem in each step of this Hegelarian ladder critically discussed by Kaufmann.[26] However, this process has several shortcomings that a dialectical image and following on the dialectical landscape can overcome:

- When posing the problem there might not be the apparent polarity of solutions. There might be several options that go beyond the pure A and not A. It can oversimplify matters or introduce biases that would guide the true solution. Especially when applying the modern western scientific methodology that aims at reduction in order to gain a generalised hypothesis that can be repeatedly proven or disproven in a particular contexts and at any time achieving always the same answer.

- The concept of a final truth is problematic in itself. Is it at all present? Could there be many different truths out there, one for each individual? Might the true solution of a problem depend upon time and develop into something different as time passes by? Is the solution to the dialectic discourse always a step into the right direction to discover the solution to a problem? Bear in mind the different usages of many groups of many sites of archaeo-astronomical interest; each claiming to have uncovered the real meaning or truth of this site. For example, Stonehenge offering a place of public worship for the Druids or an innovative and alternative venue for a solstice event for an estimated 12,500 visitors that had a mixed interpretation of Stonehenge.[27] Other usages might include stone circles such as The Nine Ladies on Stanton Moor as examples of cultural heritage and linking its value to the dark sky agenda, utilising its perceived energy to support meditation and healing, or using it as a symbol for environmental

---

26 Walter Arnold Kaufmann, *Hegel: A Reinterpretation* (South Bend: University of Notre Dame Press, 1978), Ch. 37.

27 Robert J. Wallis and Jenny Blain, 'Sites, Sacredness, and Stories: Interactions of Archaeology and Contemporary Paganism', *Folklore*, Vol. 114, no. 3 (2003): p. 309.

damage caused through quarrying.[28] This is true even in the scientific community where hypothesis offer a hiding place for the constraints forced upon us by a certain methodology. In terms of a true methodology this can be seen in the derogative critique of Fleming who describes the work inspired by, as he terms it, hyper-interpretive narratives and phenomenology as 'rhetoric and the uncritical acceptance of the output of sometimes fevered imaginations'.[29] Especially the phenomenological landscape approach proposed by Tilley when analysing the megalithic monuments of the southwest of Wales in *Phenomenology of the Landscape* triggered his response, since it was seen to be going beyond the evidence, identifying more than one significant landscape feature important for a monument's location, and overall attempting a more general interpretation.[30] Tilley responds by indicating that Fleming never engaged with the philosophical, theoretical, or conceptual issues of the phenomenology applied to landscape. Especially illustrative toward the concept of a single truth is Tilley's statement regarding Fleming's criticism that 'instead of a nuanced and multifaceted perspective on landscape, a simple black-and-white perspective in which there can only be one reason why a specific location might have significance rather than many' is used and '[Tilley's] approach becomes constituted as Other [and amounting] to a parable of good versus evil'.[31]

---

28 Daniel Brown, Fabio Silva and Rosa Doran, 'Archaeo-Astronomy and Education', *Anthropological Notebooks 19 (Supplement)*, (2013): pp. 518; Kal, Nine Ladies P2—A Healing Ceremony at http://www.hedgedruid.com/2008/09/nine-ladies-p2-a-healing-ceremony/ [Accessed 12 May 2014]; John Vidal, 'Protesters Dig in to Save Landscape from Quarry: "It'll cost millions to get us out"', *Guardian*, 14 February 2004, available at http://www.theguardian.com/environment/2004/feb/14/activists.conservation [accessed 12 May 2004].

29 Andrew Fleming, 'Post-processual Landscape Archaeology: A Critique', *Cambridge Archaeological Journal*, Vol. 16, no. 3, (2006): p. 279.

30 Christopher Tilley, *Interpreting Landscapes: Geologies, Topographies, Identities; Explorations in Landscape Phenomenology 3* (Left Coast Press, 2010), pp. 471–81.

31 Ibid., p. 478; ibid., p. 481.

- The process of arriving at the final truth is defined by a clear and linear combination of dialectic discourses. One has to follow them in this exact order predefined by a previous individual. Here this individual has power over guiding you to their real truth. When dealing with an alignment this might be seen as the initiated astronomy cast of an ancient society. Only the initiated would be aware of this only true path of meaning.

The theoretical concept of the dialectical image—a general concept from which the dialectical landscape will follow, defined by the author in the next paragraph—was proposed and described by Benjamin throughout his correspondence with Adorno and stated by Auerbach.[32] It is an attempt at solving these problems. It breaks with the linear approach of reasoning outlined above by offering the possibility to negotiating a (hypothetical) image constructed through a constellation of moments. Whereby, several possible arguments, facts, impressions, feelings, and moments in general are perceived in unity for example on a two dimensional surface of a painting. There is no path to follow, no pre-defined overlaid structure similar to a mind map, and even the concept that a single object is perceived on its own does not hold. In a painting the viewer encounters everything at once. As a consequence, the image allows an individual to see all possible solutions and factors relevant to him at the same time in an interconnected manner or constellation. A path can be woven and negotiated in any way. The whole idea of presenting it without a sequence allows seeing things in context and seeing indeed the bigger picture. The constellation of moments might be different for each and every one. Removing the constraints imposed by structures of power will in Benjamin's words offer a 'revolutionary chance in the fight for the oppressed past' and empower the individual.[33] The moments as part of a constellation are never time independent conclusions of an individual, but fleeting, come and go, or remain as more significant for an individual

---

32 Walter Benjamin, *On the Concept of History*, (Classic Books America New York, 2009); Walter Benjamin, 'The Correspondence of Walter Benjamin, 1910–1940', Gershom Scholem and Theodor W. Adorno (eds.), trans. Manfred R. Jacobson and Evelyn M. Jacobson (Chicago: University of Chicago Press, 1994); Anthony Auerbach, 'Imagine No Metaphors: The Dialectical Image of Walter Benjamin', *Image & Narrative*, Vol. 18 (2007).

33 Benjamin, *On the Concept of History*, p. 396; Tilley, *A Phenomenology of Landscape*, p. 40.

16 The Experience of Watching

making this dialectical image. The constellation will alter adapting to the times and changes present in both the individual as well as society.

Applying this theoretical concept to a landscape creates a dialectical landscape. Here the described moments can be identified to some extent by objects seen in the landscape that symbolise certain characteristics of a landscape. For example a quarry might represent the natural resources defining a landscape. But moments are entities in time and are far more than physical objects. The before mentioned quarry might be overgrown and represent a current status of mining in the region or cause the viewer to feel delighted how nature has started to reclaim this patch of landscape as it might have done in many instances before. The moments are meaning and emotions unlocked in the rhythmic flow of time. They are part of the package a viewer of the landscape has brought with them as memories. The viewer's baggage manifests itself in a landscape that acts as a pointer towards times gone by. The image is only used to visualise the concept. It contains far more than just visual stimuli: touch, smell and hearing. All senses come together to present an array of moments loaded with emotions and causing emotions by themselves.

Applications in a Landscape
At this point it should be noted that reading this article in the seclusion of one's study or listening to it being read in a grey concrete lecture theatre might not be the best environment to follow the ideas presented here. It would be a far better option to walk or, even better, wander through your local neighbourhood or countryside. Such drifting and becoming lost is an approach described by Coverley as psychogeography.[34] One becomes a flâneur, feeling the personal meaning of the landscape and how places have a distinct emotional impact on us.[35] To capture this notion the author would like to present a brief description of what a view into the landscape gathered at a frequently visited location might offer. This location was chosen since it is part of a landscape visited in many different contexts as well as to develop the author's approach to phenomenology. It is lacking in covering all the aspects raised above but should be seen as a starting point for the reader's own exploration and wanderings. It also forces the reader to follow the author's winding trail across the picture with moments

---

34 Merlin Coverley, *Psychogeography* (Oldcastle Books Ltd, 2010), pp. 9–30.

35 Walter Benjamin, Harry Zohn and Q. Hoare, *Charles Baudelaire: A Lyric Poet in the Era of High Capitalism* (London, 1973), p. 54.

inspired through the presence of many aspects and emotions simultaneously. All this cannot be describe in a comprehensive narrative but should encourage the reader to go and experience the dialectical landscape themselves and see where their journey leads them.

Figure 1 Views into the Derwent Valley at Gardom's Edge, Peak District, UK. The image contains many physical objects possibly representing power and time as well as triggered emotional responses.

The imaginary journey of the author takes place in a view captured in parts in Figure 1. The image shows the landscape at Gardom's Edge in the Peak District as a ridge containing a large amount of cultural heritage:

> *'Standing at this point I gather the first view into the Derwent valley and the expansive horizon framing the rolling clouds. In the distance I can note a disused quarry illustrating how we have mined this region to access its resources. Such activities have left behind scars in the landscape that are slowly recovering. The mining itself did not take into account the interaction with nature nor was it sustainable. As such it acts as a reminder how [humans] can destroy nature. However in close proximity I can notice two rectangular fields where one can just make out the letters E and R. These are a plantation created to honour the Queen. They are a mark in the landscape clearly visible for miles and stating a cultural aspect of our current society in a much more sustainable fashion. Both mining and*

commemorative plantations are solutions to how humans interact with nature in recent times, and specifically in context of colonisation and rehabilitation.[36]

In the foreground of the image two muddy paths are visible, one of which is an ancient track used centuries ago prior to the old toll road below in the valley. If I follow the path down the ridge entering the denser woodland, a Victorian Smithy awaits, moss overgrown and in the deep cool shadows of the trees. Surely something one might overlook when walking along this route. Here I cast my mind back at several school classes guided down this route and their explorations of the surrounding landscape. But taking the path up the ridge will lead along a much more recent path used by some walkers to enter Gardom's Edge and Birchen Edge. This fork represents the entrance onto the ridge and also a short respite while walking up the incline.

The junction itself is marked by a rather large rock. This is sedimentary sandstone, Millstone Grit, the bones of this landscape, shaping and forming its outcrops and slopes. This rock reveals what was here millions of years ago and opens up our perception of time exploration on a much larger scale. It also was the starting point for my journey towards geology and analysing sedimentary sandstone, a topic quite far removed from my usual work. The two paths are both recent and centuries old created through the human rhythmic task of walking. The rock represents times before humans were part of this landscape. The distant views reveal disused quarries and plantations that are both recent results of human interaction with the landscape: one fading away slowly but surely as its creative rhythm dies down while the other reveals powers present in our society through its continuous up keeping.

At this time and location I can also draw upon my own experience of time: how this was the point at which dawn revealed a rising mist and the still silent main road in the valley at an equinox visit to the site after having ventured up hill in darkness; how the dense falling snow inhibited any views beyond 10 meters and forced the group of international school teachers to turn back down towards the shelter of the inviting Inn at the car park. Such musings might then lead on to me thinking: we have picked a good time to come here.

Again, the train of thought might be drawn back towards our impact on nature and sustainability or light pollution. Frequently this offers time for a discussion with students that accompany me on this walk, what conservation of this untouched moorland really means.

---

36 Gruenewald, *The Best of Both Worlds*, p. 4.

All these different aspects of the landscape as well as emotions and memories that appeared in context might illustrate how the moorland itself is not the untouched landscape we might imagine but is an ever changing environment adapting and incorporating human activity.[37]

During these musings it has become apparent how the landscape itself triggered emotions and feelings. The viewer within the landscape when opening all their senses turns into a subject upon which the landscape with all it facets can act upon. When engaging with the dialectical landscape the viewer hears the voice of this place as described by Lane.[38] Walking in the landscape described briefly above wants the viewer to move on and see more. Walking becomes exploring and learning more about where one is.[39] The landscape triggers action even in the most basic way of walking a path and being part of the rhythmic cycle giving live to a network of paths. The experiences call for action, they might motivate the walker to bring along friends or trigger the urge within locals to protect dark skies stretching above their landscapes full of cultural heritage. Suddenly the viewer has become an object to act upon the subject of landscape; to leave a mark on the stage presented by the landscape similar to Cosgrove and Daniels idea of a landscape being a collection of material objects on a pre-existing stage or symbols represented on a provided canvas or backdrop.[40]

But realising that the viewer or landscape are both subject and object within time resolves the difficult concept that Lane argues as '...the place perceiving itself through us' and can actually take part in a communication

---

37 Daniel Brown, Esther Johnson and Mary Brittain, (2012), 'Work Experience at Nottingham Trent University', at http://www2.warwick.ac.uk/fac/soc/cei/centrelinkmagazine/marchcontents/workexperienceatnottinghamtrentuniversity/ [Accessed 28 March 2014].

38 Belden C. Lane, 'Giving Voice to Place: Three Models for Understanding American Sacred Space', *Religion & American Culture*, Vol. 11, no. 1, (2001): p. 58.

39 Daniel Brown, Natasha Neale, and Robert Francis, 'Peak into the Past—An Archaeo-astronomy Summer School', *School Science Review*, Vol. 93, no. 342, (2011): p. 83; Daniel Brown and Lina Canas, 'Archaeoastronomy in Society: Supporting Citizenship in Schools Across Europe', *International Journal for Science in Society*, Vol. 2, no. 3, (2010): p. 155.

40 Cosgrove and Daniels, *The Iconography of Landscape*, p. 1.

with the viewer.[41] Perceiving landscape purely as a collection of physical objects even if created by humans in the past or present does not allow an actual discussion with a landscape. Only by realising that one is part of the landscape through one's own actions and emotions over time (expressed in Greek as *chronos*) will it in this unrepeatable moment in time full of significance (expressed in Greek as *kairos*) become apparent that the landscape can indeed speak and listen, a voice and ear given to it by those who dwell within it over time becoming part of the landscape. And as Lane points out, such an 'experience (…) simultaneously in a situation of *chora* (place) and a moment of *kairos* is truly to encounter wonder'.[42] This could even be focused towards the instance of experiencing place through the realisation how *chronos* and *kairos* come together and resemble one of the true moments of Benjamin's dialectical image described by Auerbach.[43] Such a realisation triggered by a dialectical landscape that has a constellation of moments at its heart becomes a new moment in itself. The new moment can form a constellation of moments gathered while watching and walking in a landscape. This results in the above mentioned transient and highly temporal experience of the constellation of moments but also goes beyond a place experience towards describing a landscape as a constellation of moments.

Seascape and More
Landscape is as fluid a term as place and contains many physical and non-physical properties and objects within, best expressed by experiencing it rather than by geographic location or outlines. So far the author has been using landscape as the general term collecting all the other scapes derived from it such as seascape, skyscape and landscape. The latter is initially defined in the traditional way. All three together were first brought to a critical attention through an art installation by Dobrowolski and Painter.[44] Looking at one part of this trio of scapes, the seascape, will allow us to

---

41 Tilley, *Interpreting Landscapes,* p. 40; Lane, *Giving Voice to Place*, p. 58.

42 Lane, *Giving Voice to Place*, p. 55.

43 Auerbach, *Imagine No Metaphors.*

44 C. Dobrowolski, C. Painter, & Ferens Art Gallery (2002) *Landscape, Seascape, Skyscape, Escape.* Kingston Upon Hull City Museums, Art Galleries & Archives.

outline in more depth how to explore place and finally that this trio is indeed only one.

The foundations of a seascape were laid out by Westerdahl stressing that the sea should not be treated as a limit at which cultures and trading end to reappear at some other part of the coast, but that maritime cultural landscape is needed and the sea itself can be a social arena.[45] The term seascape was then introduced by authors such as Wehlin.[46] Standing at the beach or looking over the sea from a boat it becomes clear that this scape is completely different than the traditional landscape: One cannot actually see any objects at first sight at high sea. A first step to exploring this place within a seascape would be to look below the surface to track the geography of the seabed, locate ocean current and mark old wrecks. Here the viewer literally has to look at what lies below. However they are here brought to this place because of frequented trading routes across sea. It might also coincide with a promising fishing ground determined by the rhythmic coming and going of marine wild life over the seasons. And at this stage one has immersed one selves in the ebb and flow of time to explore place. What seems so difficult to integrate in the perception of place in a landscape becomes a necessity for a place in a seascape ruled in many cases by the tides.

Noting how the water actually obscures many aspects of what makes a seascape, it is only a small step to note that the Sun with its light during the day acts in a similar way through obscuring a skyscape visible at night time. Like the sea rolling in with each wave covering and revealing the beach that has itself been uncovered by the outgoing tides, the rhythmic sunrise and sunset can be compared to tides giving a skyscape an added dimension. Exploring this new aspect calls for a process of observing that embraces time or more appropriately, watching. Discovering such cyclically revealing aspects of a skyscape assists in understanding meaning contained within them especially expressed in concepts such as heliacal rising, the first visibility of a star after being invisible for a period of days in the morning sky before it is washed out by the incoming tide of sunlight. Discovering such a metaphorical link described above between seascape and skyscape occurred at the physical boundary of where land and sea

---

45 Christer Westerdahl, 'The Maritime Cultural Landscape', *International Journal of Nautical Archaeology*, Vol. 21, no. 1, (1992): p. 6.

46 Joakim Wehlin, 'Approaching the Gotlandic Bronze Age from Sea: Future Possibilities from a Maritime Perspective', *Gothland University Press*, Vol. 5 (2010): p. 89.

22 The Experience of Watching

meet, a beach. The importance of such boundaries, for example as outlined by Wehlin, has been marked by societies not only through noting tides but also marking their landfall at certain places and marking places with monuments. There are further boundaries occurring in this trio of scapes.[47] The immaterial mathematical and astronomical concept of a perfect horizon can be defined by where sea and sky meet. This defines when an object will rise or set in a location with a theoretical mathematical horizon most closely resembled by the sea level. This rising or setting point is a virtual place that can never be reached or a direction that can be followed. The other boundary is defined by land and sky forming the real horizon. A star setting above a mountain seen along an alignment within a monument is a real location that can in fact be visited and explored.

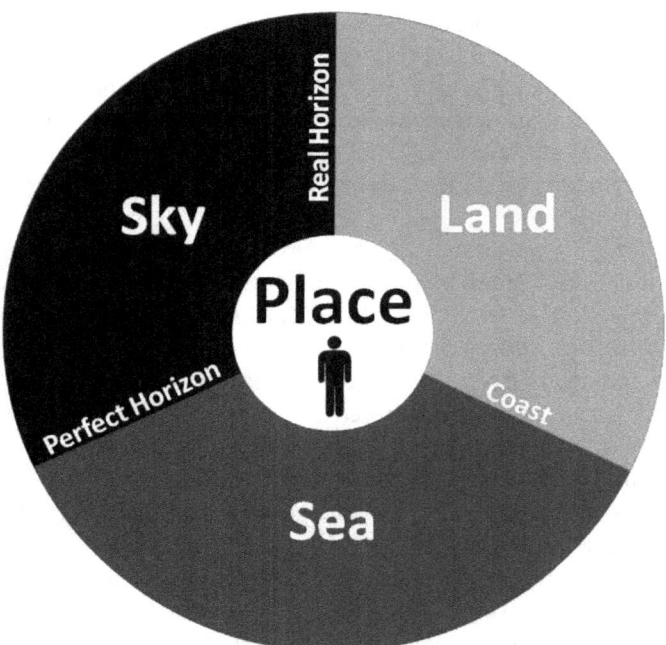

Figure 2 The trinity of sky, land and sea representing the unity of landscape when an individual discovers place and himself in all components.

But such boundaries triggering metaphorical musings and offering moments within a dialectical landscape might not always be described by

---

47 Ibid., p. 103.

the above mentioned physical boundaries. Recalling the experiential nature of all three scapes it can be seen that they can meet in many other places offering us unforeseen insights. The stars or Moon reflected in the surface of a lake bring them down to Earth and turns them into object of our world.[48] Depicting constellations seen in the sky, our societies frequently create their own land- and seascapes claiming ownership of the stars above. Viewing a valley and villages therein after night has fallen, reveals a plethora of lights weaving their patterns across the land similar to stars in the sky. By removing their own ability to view the constellations, they have themselves become a constellation of light.

Having briefly strayed into possible moments of realisation, these thoughts illustrate that the artificially created components of land-, sea- and skyscape are indeed a trinity and can only be fully understood and explored when experiencing their unity as illustrated in Figure 2. This cannot occur through a scientific approach excluding as it must any emotional response, for example through observation. The exploration has to embrace the task of watching and experiencing sea, land and the sky over time. At this point, rather than describing this trinity as a landscape it has become a place in which one dwells. This realisation occurs at the moment when the human viewer realises that they are not a fourth component needed to add to the trinity, but they are already part of the land, sea, and sky around them which is well illustrated by night time landscape installations.[49] They and societies past and present are part of something bigger through their actions in time, as Tilley claims '...a pattern of activities 'collapsed' into an array of features'.[50] This realisation is open to everyone and reveals itself to the individual through watching. This stands in stark contrast to one of Tilley's ideas that a landscape 'cannot be understood just by introspection [and t]heir meanings and significance must be taught by some and learned by others' compared to an individual that has to encounter revelation through their own individual journey.[51]

---

48 Bernadette Brady, *Dragons, Mountains and the Sky*. Presentation given at the 2013 meeting of the Theoretical Archaeology Group. University of Bournemouth, 2013.

49 Nina J. Morris, 'Night Walking: Darkness and Sensory Perception in a Night-time Landscape Installation', *Cultural Geographies*, Vol. 18, no. 3, (2011): p. 334.

50 Tilley, *A Phenomenology of Landscape*, p. 162.

51 Tilley, *Interpreting Landscapes,* p. 40.

This journey is not guided by a teacher but by the learner's own emotions and memories.

Conclusion
The challenge of modern archaeoastronomy in its search to discover the meaning behind alignments has been outlined. Statistical methods and pure scientific approaches can only help to a limited degree. Only by exploring the site of an alignment in situ within the context of the entire landscape including sea, land and sky can place at which the alignment was found be experienced. The concept of seascape has been used to illustrate the inherent interconnectedness of all three scapes. The place experience is supported by the process of watching and becoming aware of the passage of time, the rhythms and cycles apparent in one's surrounding. The time experience opens up an ability to emotionally connect and communicate with a landscape; at this point one truly dwells there. The concept of the dialectical landscape has been outlined as a way for the viewer to negotiate a journey towards finding themselves in a landscape. In doing so, the viewer will note that time has made them part of it and its cosmic cycles.

It can be concluded that exploring one's own reflection as well as the reflection of societies and humans in the landscape makes us become part of something bigger. A meaning for an alignment can now be negotiated within the landscape with emphasis on its skyscape component that was first made significant by statistical methods. The holistic approach of watching will lead us on to listen to the voice of the landscape that was both witness and motivator to the erection of both monument and alignment. Although talking to the builders of an alignment is not possible their voices still echo in the landscape waiting to reveal its meaning. As outlined the dialectical landscape rejects any one truth or solution to a problem, acknowledging the importance of the individual. However, the contemporary viewer is most commonly a member of society influenced by the western scientific methodology. Therefore, it might appear fruitless to disentangle the voices of the builders of an alignment within a landscape entangled in many others voices over the millennia. Furthermore, the task becomes more challenging when heard by someone not understanding the language. But the author is confident that this is the right way to proceed, since this approach now offers a metaphorical text that can be utilised to translate and interpret, rather than a method which is limited by statistical methodology surrounded by a void of uncertainty.

Skyscapes of the Mesolithic/Neolithic Transition in Western England

Pamela Armstrong

Abstract:
This paper explores the skyscapes of central southern England during the Mesolithic/Neolithic transition. It suggests those who built the structures known as Cotswold-Severn tombs embedded architectural details within their monuments that linked to celestial horizon events for both navigational and calendrical purposes. The architectural designs found within the tombs are analysed for evidence of a cultural engagement with landscape archaeology and the sky. This period witnessed a transition between two eras, so this research considers the possibility that there may also have been a shift in the type of skywatching practised by those inhabiting this landscape during this time of change.

Introduction

The question considered for this paper was, 'Does the archaeoastronomic record of the Cotswold-Severn region reflect evidence of a transition from lunar to solar alignment?' The monuments surveyed for this research were Neolithic Cotswold Severn long barrows, much like the one below— known as Crippets (Fig. 1). These are earthen mounds which have been described by Kinnes as 'the finest group of stone chambered tombs in England'.[1]

---

[1] I Kinnes, R J Mercer, and I F Smith, 'Research Priorities in the British Neolithic', (unpublished report submitted to the DoE by the Prehistoric Society, 1976), p. 6.

Figure 1 Crippets long barrow. Cotswold escarpment. Gloucestershire UK. Own photo.

North defines a barrow as 'a mound, deliberately erected out of earth and other material [...] and having a conscious architectural structure. Usually, but not always, built for burial purposes'.[2] Darvill estimates that there are about 500 long barrows in Britain.[3] The Cotswold Severn tombs began to appear on the southern English landscape at the onset of the fourth millennium BCE and currently about 200 barrows have been recorded in the region.[4]

The idea for my research grew from Sims' suggestion that Stonehenge is a Neolithic monument designed by those who built it to 'juxtapose, replicate and reverse' key horizon properties displayed by the sun and moon, apparently in order to invest the sun with the moon's former

---

2 John North, *Stonehenge Neolithic Man and Cosmos* (London: Harper Collins, 1997), p. xxiii.

3 Timothy Darvill, *Long Barrows of the Cotswolds and Surrounding Areas* (Brimscombe: Gloucestershire 2004), p.71.

4 Darvill, *Cotswolds*, p. 83.

religious significance.⁵ He describes this process as 'solarization'.⁶ Sims argues that prior to the shift to pastoralism at the beginning of the Neolithic, there was a greater cultural engagement with lunar rather than solar astronomy. Sims suggests that pre-Neolithic, communities organised themselves by 'phase-locking their economic and ritual routines to the rhythms of the Moon'.⁷ He claims that Stonehenge's sarsen stone circle was used to 'modify and transcend' a lunar allegiance by superseding it with a solar one.⁸ Thus it was in his view, a mechanism for engineering social change. It was essentially a binary monument, structurally designed to facilitate a symbolic transposition from the moon to the sun. Sims suggests his hypothesis be tested by investigating the region surrounding Stonehenge, which my research endeavoured to do. Most specifically I was exploring whether it is possible to identify a continuity or discontinuity of astronomic allegiance to luminaries across the Mesolithic to Neolithic transition.

A literature review quickly establishes that the barrows have a singular feature, and that is simply their massive bulk. The tombs were new to the landscape. Thomas describes them as 'perhaps the most archaeologically conspicuous element of the British Neolithic'. Indeed they mark the onset of this period, Darvill suggesting the new use of megaliths indicated 'a step-change' in construction and design.⁹ As DeMarrais points out:

> Monuments can be impressive, even overwhelming constructions that are experienced simultaneously by a large audience. They are an effective and enduring means of communication, often expressing relatively unambiguous messages of power.¹⁰

---

5 Lionel Sims, 'The 'Solarization' of the Moon: Manipulated Knowledge at Stonehenge', *Cambridge Archaeological Journal*, Vol. 16, no. 2, (2006): p. 1.

6 Ibid., p. 2.

7 Ibid., p. 3.

8 Ibid.

9 Darvill, *Cotswolds*, p. 71.

10 E. DeMarrais, L.J. Castillo & T. Earle, 'Ideology, Materialization, and Power Strategies', *Current Anthropology*, Vol. 37, (1996): p. 18.

Scarre argues that by building such monuments, the earliest Neolithic communities 'established a pattern of behaviour that set them apart from their Mesolithic antecedents'.[11] So the earthen long barrows were radical in appearance, but there is debate about who built them. When Thomas writes of the social complexity of the 'Neolithic transition' he describes the phrase itself as 'protean'.[12] In his view, it is a term with meanings ranging from:

> a particular type of subsistence economy, a level of technological development, a chronological interval, a specific set of cultural entities, to racial or ethnic identities, or to a specific type of society.[13]

A significant feature of the shift from the Mesolithic was the move from a predominantly forager way of life to a generally agrarian one.

In some places the Mesolithic to Neolithic transition appeared to spread gradually across a region, in other places it is sudden. Rowley-Conwy contends that the appearance of agriculture was 'not a demic "wave of advance" but rather a rapid and massive socioeconomic "wave of disruption."'[14] If he is correct, then monument building and its associated belief systems may have been in response to the tensions caused by that 'disruption'. So these monuments emerged at a time of great change though there can only be speculation as to whether the ideologies associated with them reflected or indeed shaped inherent societal shifts. If they contained features to do with astronomy in particular, then this astronomy was either new to the landscape or it was being embedded in the material culture in a new way. Ruggles suggests this can be process which generates tension, writing:

---

11 Chris Scarre, 'Changing Places: Monuments and the Neolithic Transition in Western France', in *Going Over: The Mesolithic-Neolithic Transition in North-West Europe*, eds. A Whittle and V Cummings (Oxford: Oxford University Press, 2002), p. 243.

12 Julian Thomas, *The Birth of Neolithic Britain* (Oxford: Oxford University Press, 2013), p. 1.

13 Ibid.

14 Peter Rowley-Conwy, 'How the West Was Lost. A Reconsideration of Agricultural Origins in Britain, Ireland, and Southern Scandinavia', *Current Anthropology*, Vol. 45, (2004): p. 97.

discontinuities of ritual tradition, as manifested by clear changes in the patterns of astronomical symbolism incorporated in public monuments, may indicate significant social upheaval.[15]

This would support Thomas' view that the barrow builders had new and pressing economic imperatives. He writes:

> People do not bury themselves: the burial of the dead is an aspect of the power strategies of the living. These new burial traditions were a means by which the inheritance of land and wealth from one individual to another was made legitimate.[16]

Thus the Cotswold Severn earthen tombs may have functioned as a statement of intent, built to establish lineage and ownership in what was possibly a contested environment.

On the other hand it is possible there were regions where the transition did not precipitate upheaval, in which case monuments may have fulfilled altogether different functions. Silva and Franklyn point out that when divergent populations interact such as may have occurred between Mesolithic and Neolithic peoples, it is possible for transformation to occur rather than conflict. When considering belief systems for instance, they suggest that different communities each could act on each other, creating 'syncretic cosmologies with elements from both the colonised and coloniser's world-views'.[17] Thus, the emergence of the barrows in western England may have fulfilled a complex and not necessarily uniform suite of needs. What is known is that they were usually places of burial and thus memorial. Rowley-Conwy suggests they were monuments which constituted:

---

15 Clive Ruggles, *Astronomy in Prehistoric Britain and Ireland* (New Haven: Yale University Press, 1999), p. 152.

16 Julian Thomas, 'Relations of Production and Social Change in the Neolithic of North-West Europe', *Royal Anthropological Institute of Great Britain and Ireland*, Vol. 22, no. 3 (1987): p. 423.

17 Fabio Silva and Roslyn M Frank, 'Deconstructing the Neolithic Myth: The Implications of Continuity for European Late Prehistory', *Anthropological Notebooks*, Vol. 19, no. Supplement (2013): p. 232.

a durable focus for a community, enhanced by the physical presence of one or more founding ancestors who served to emphasize continuity with the collective past.'[18]

But Thomas points out that even burial, a defining task universal to humanity, differed after the transition into the Neolithic. The people of the Mesolithic did not build enduring tombs, whereas in the Neolithic, he argues, monument building became 'a social strategy, which had the effect of bringing people together to labour and to engage in ritual observances'.[19] The very act of monument construction, he suggests, was a communally cohering event. Thus a monument was a public statement, imbued with social meaning.

In the case of the earthen tombs they may have been intended not just as markers establishing territorial boundaries, but may have been used to enshrine shared cosmologies. Ruggles agrees that the tombs manifested a social and political complexity. But he is amongst those who also suggest they had astronomic features. He says of the barrows that their 'orientation was certainly important'.[20] He also writes:

> Perhaps, in small communities, astronomical alignments simply helped to affirm a monument as being at 'the centre of the world', but in other cases they may have had more to do with making its power impossible to challenge thereby affirming ideological structures and political control.[21]

As DeMarrais points out, the design of a structure is 'a means through which symbols, their meanings and beliefs can be manipulated to become an important source of social power'.[22] Whatever the function of the barrows, whether they were built by indigenous Mesolithic hunter gatherers, incoming Neolithic farmers or by selective appropriation between the two, this new architecture heralded the end of the Mesolithic

---

18 Thomas, *Neolithic Britain*, p. 92.

19 Julian Thomas, 'Thoughts on The "Repacked" Neolithic Revolution', *Antiquity*, Vol. 77, (2003): p. 72.

20 Ruggles, *Astronomy in Prehistoric Britain and Ireland*, p. 125.

21 Ibid., p. 154.

22 DeMarrais, '*Power*', p. 31.

in western England, and it may have contained astronomic intent. When a culture embeds the astronomy it practices within the fabric of a new structure, it is a declarative act inferring that continuity will apply. For those who are establishing territorial or ideological boundaries, an intended alignment from a power base—such as a barrow—to a celestial event links past, present and, critically, the future.

The methodology used in this research was shaped in part by the limited material record available. As Whittle points out, 'only three Cotswold long barrows or cairns have been more or less fully excavated'.[23] Because of this I decided on a case study approach which focused on the three barrows Whittle specifically nominates. These were Burn Ground, Ascott-under-Wychwood and Hazleton North (Fig. 2).[24]

Figure 2 Barrows' North Cotswold location. Google earth image.

The value of a case study approach is that, as Stake suggests, it gives the researcher the opportunity to get to know each case 'extensively and

---

23 Alasdair Whittle and Don Benson, 'Place and Time: Building and Remembrance', in *Building Memories the Neolithic Cotswold Long Barrow at Ascott-under-Wychwood, Oxfordshire* (Oxford: Oxbow, 2007), p. 327.

24 Benson, 'Building and Remembrance', p. 327.

intensively'.²⁵ The case study approach is qualitative, and differs from quantitative research because, argues Stake, it 'seeks out a relationship between a small number of variables'.²⁶ Quantitative research looks for patterns amongst a large number of research objects. It is reductive and therefore difficult to apply to the Cotswold-Severn barrows, whose designs are complex and varied. Each barrow is different. Though there may be broad commonalities, no one design is commensurate with another, and it is impossible to reduce their architectural features to a manageably small set of significant markers. As Timothy Darvill explains, the barrows are monuments that display 'very considerable heterogeneity'.²⁷ A quantitative approach militates against assessing the varied details particular to each barrow so because of this, a qualitative approach was chosen.

When I mentioned the three barrows above I used the past tense. The reason for this is that the very process of full excavation completely destroyed them. This total lack of material record has led to a reconsideration of what, in relation to this study, constitute primary and secondary sources. Given Benson's confident assessment of the archaeological reports which record and describe the three excavations, I made those documents my primary material.²⁸ Thus my primary sources became not the barrows themselves, but the reports associated with them.

The second reason that I chose to focus on these reports was because they provided dependable dates. This dating process allowed for the establishment of a time frame within which to compare and contrast each barrow. Burn Ground was built possibly at the end of the fifth millennium, around 4230–3970 cal BCE.²⁹ Ascott-under-Wychwood was built just after the beginning of the fourth millennium, around 3760–3700 cal BCE.³⁰ Hazleton North followed immediately, around 3710–3655 cal BCE.³¹ Thus

---

25 Robert E Stake, *The Nature of Qualitative Research* (London: Routledge, 1995), p. 36.

26 Ibid., p. 41.

27 Darvill, *Cotswolds*, p. 44.

28 Benson, 'Building and Remembrance', p. 327.

29 Brickley, 'Date and Sequence of Use', p. 339.

30 Benson, *Excavations*, p. 226.

31 Meadows, Barclay, and Bayliss, 'Dating of the Hazleton Long Cairn', p. 54.

these tombs predate Sarsen Stonehenge and, if it occurred, the 'solarization' period, by anything up to 1,500 years.[32]

Given the above, my methodology became a hybrid one which included the combined use of archaeological reports, maps, and illustrations. Fieldwork calculations, phenomenological notes and a discussion of the horizon issues local to each site were also included. The calculations I arrived at combined fieldwork and map work. Modern features such as roads were used as key structures. I would locate the road nearest to a barrow and go into the field and measure its azimuth. I then used maps and archaeological plans to draw the angle between the road and the barrow. A protractor was used to calculate the difference. This hybrid approach was more dependable, the more precise the maps and illustrations. It was a methodology which had its limitations and was less reliable when a badly, or roughly drawn diagram was used. This deductive process and the declinations which result from it, are fully discussed in my end summary.

The tools used for field work included a Garmin GPS 12 XL position finder, as well as a Suunto compass. A Suunto clinometer was used to measure horizon altitude. Magnetic anomalies were checked for at the location of all three excavations. Photographs and archaeological diagrams were used to infer the barrow's location as best possible. When checking for magnetic anomalies, two poles were inserted into the ground along the most probable orientation for each barrow and a compass was used to check the azimuth in each direction. None were found. True North was recalculated from Magnetic North by accessing the National Geophysical Data Centre's website.[33] As this research used secondary sources to impute primary source measurements, all calculations would benefit from some leeway thus an error margin of up to 2° has been used throughout.

Though my study focused on the sun and the moon, I routinely checked for stellar orientation. Thus I noted Schaefer's discussion about the 'uncertainty' of a star's extinction angle, that is, the lowest point on the horizon at which it's visible.[34] Two astronomy programmes were used: the first was Stellarium, and the second was Starlight, whose star catalogue I

---

32 Sims, 'Solarization', p. 2.

33 National Geographic Data Centre, 'www.Ngdc.Noaa.Gov/Geomagmodels', [Accessed 15 March 2013].

34 Bradley E Schaefer, 'Atmospheric Extinction Effects on Stellar Alignments', *Archaeoastronomy* 10, no. xvii (1986), p. 41.

accessed.[35] Starlight's catalogue is compiled from the Yale Bright Star Catalogue and Ptolemy's Almagest. I have restricted stars chosen to those of a visual magnitude of 3 or less. All horizons east, west, north and south were assessed for celestial events.

The remit of this research was to search for orientations from the barrows to celestial horizon events and then to judge whether these indicated a shift from a lunar to solar allegiance over time. The hybrid methodology just described allowed me to establish a diachronic profile of one small part of the material record of the Mesolithic to Neolithic transition on the landscape just north of Stonehenge.

Earthen barrows contain both internal and external architectural features, and orientation can be measured from both. In general, barrows fulfilled a mortuary function, earthen mounds generally subsuming stone chambers containing collections of bones. Jenni Anderson has kindly permitted use of her illustration of the foundations of a Neolithic mortuary structure which later became a barrow (Fig. 3).

Visualisation of the deposition of human remains within the first mortuary structure at Wayland's Smithy 1, according to Ian Kinnes' published theory, 1975. Watercolour & crayon on watercolour paper, 2009.

Figure 3 Artist Jenni Anderson's depiction of the primary tomb at Waylands Smithy.

Anderson's drawing interprets Kinnes' description of his Wayland's Smithy excavation in 1975. The structure was assumed to be a ritual space

---

35 Stellarium 0.12.0; *Starlight*, www.Zyntara.com.

for the dead.[36] The lithics forming the small mound at the centre of the illustration were the stones found at the heart of the barrow. The barrow would only have come into existence when the stones were covered with earth and even more stones. And it is at this point that a choice is made. The new structure could be round or long. It is of interest to the archaeoastronomer that the form chosen for the Cotswold-Severn barrows was an elongated one which laid an axis across the landscape. As Darvill points out, 'since one essential feature of a long barrow is its linear form, each will naturally have an orientation.'[37] Where there is an orientation, there may be a deliberate alignment to a celestial horizon event. Certainly this is the case with the Cotswold-Severn long barrows, which offer such a rich resource for investigation of astronomic intent. The aerial photograph below shows Wayland's Smithy today, lying across the centre of a copse (Fig. 4).

Figure 4 Waylands Smithy long barrow. Google earth image.

---

36 I Kinnes, 'Monumental Function in British Neolithic Burial Practices', *World Archaeology*, Vol. 7, no. 1, (1975).

37 Darvill, *Cotswolds*, p. 97.

36 Skyscapes of the Mesolithic/Neolithic Transition in Western England

The mound below is known as the Gatcombe barrow (Fig. 5).

Figure 5 Gatcombe Long Barrow. Own photo.

One of the best-maintained barrows in the region is Belas Knap, seen below (Fig. 6).

Figure 6 Belas Knap. Google images.

Culture and Cosmos

There is debate about the way barrows should be surveyed. For the purposes of this study I measured the orientation created by the barrow's length. But measurements have been taken across barrows too. North describes the possibility of observing celestial events from just below a barrow, close to its side, 'at right angles' over the mound.[38] It seems he is inferring that the topmost spine along the barrow afforded an artificial horizon against which the rise or set of sun, moon or stars could be measured. But given the length of the barrows—indeed North mentions Burn Ground, which he describes as being 30 metres long—this creates a relatively wide arc from the observer's point of view.[39] Celestial observations may have been possible, but few barrows offer evidence of having had poles or standing stones installed along their spines as possible foresights. Given this, it is unclear how any one particular azimuth was judged as more significant than another given the extent of that arc. Thus, whenever surveying a barrow I focus on the orientation created by the length of the mound itself. This was doubly necessary for this particular survey as the barrows I was researching did not physically exist. Where they were concerned, there was no artificial horizon to look at from a ninety-degree angle to begin with.

In addition, barrows are trapezoidal in shape. If viewed lengthways, their highest point acts as a foresight and their long outline confirms orientation. These two physical characteristics are interdependent and offer an incontrovertible angle towards a clearly identifiable azimuth. These combined features cannot be exploited if the barrow is approached sideways. This appears to imply that barrows provide orientation to only two horizon points, but in fact all my sites revealed orientations at right angles to the primary measurement; however the second set came from architectural features found within the barrows themselves.

The following findings result from combining my fieldwork calculations with diagrams and maps to ascertain azimuths, from which the barrows' declinations were established.

---

38 North, *Stonehenge*, p. 124.

39 Ibid., p. 126.

38 Skyscapes of the Mesolithic/Neolithic Transition in Western England

Burn Ground

Burn Ground was the first barrow in my study. The aerial photograph below is of the barrow itself, or rather what was left of it towards the end of its excavation (Fig. 7).

Figure 7 Excavation site of Burn Ground showing vestigial remains of long barrow.[40]

Burn Ground was excavated by Grimes during World War II, probably as part of the preparation for turning its mile long field into an airstrip. My survey of this barrow immediately struck a methodological difficulty. Though the excavation itself was documented there were no photographs or maps which illustrated the barrow's setting in its wider landscape. This was one of the points when the hybrid nature of my research came into play. It was only when I sourced an RAF aerial map from 1947 that I could establish where the barrow lay in relationship to its nearby road (Fig. 8).[41] Turning then to fieldwork measurements, I was able to establish first the azimuth of that road, and from that the azimuth of the barrow (Fig. 9).

---

40 Alasdair Whittle and Don Benson, 'Place and Time: Building and Remembrance', in *Building Memories the Neolithic Cotswold Long Barrow at Ascott-under-Wychwood, Oxfordshire* (Oxford: Oxbow, 2007).

41 English Heritage, 'Aerial View of Burn Ground Taken by '82 Sqdn', in *Serial No: 3280* (Archive, 28 May 1947).

Pamela Armstrong 39

Figure 8 1947 RAF aerial map.[42]

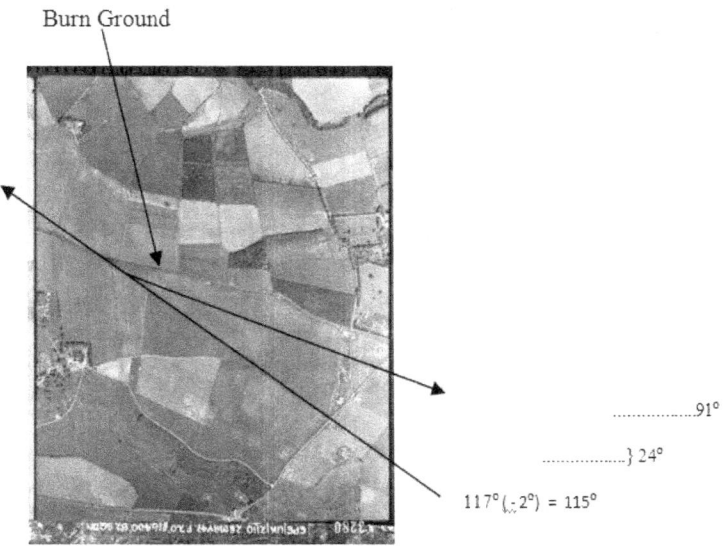
Figure 9 The azimuth of the A40, which runs along the side of Burn Ground field, and the azimuth of the barrow, in relation to it.

---

42 Ibid.

40 Skyscapes of the Mesolithic/Neolithic Transition in Western England

Grimes wrote of Burn Ground that its 'true axis was almost exactly east-west' and as can be seen, the hybrid methodology that I employed concludes the same.[43] Burn Ground may have had a single orientation, or possibly two. Zero degrees of declination orients to both the equinox and to what Silva has discovered is one of the probable rise points for eclipsing Autumn Full Moons on a minor lunar standstill.[44] The fixed stars that the barrow oriented to possibly included Alhena, Procyon, Alphard and Deneb Adige (Fieldwork Findings).

Ascott-under-Wychwood
The barrow at Ascott-under-Wychwood was excavated in 1981 (Fig. 10).

Figure 10 Excavation site of Ascott-under-Wychwood. 1981.[45]

---

43 W. F Grimes, 'Excavations on Defence Sites, 1939–1945 1: Mainly Neolithic—Bronze Age', in *Burn Ground, Hampnett, Gloucestershire* (London: Her Majesty's Stationery Office, 1960), p. 43.

44 Fabio Silva, 'Equinoctial Full Moon Models and Non-Gaussianity: Portuguese Dolmens as a Test Case', in *Astronomy and Power*, eds. Barbara Rappenglueck and Nicholas Campion (British Archaeological Reports, 2011).

45 Benson, *'Building and Remembrance'*, p. 381, Plate. 1.1.

Its declination was +9.2/-8.4°, identified by Silva as one which finds the rising Autumn Full Moon, or the rising eclipsing Autumn Full Moon at minor lunar standstill (Fig 11).[46] In terms of fixed stars, the barrow also oriented to Alcyone, Deneb Adige, Aldebaran and Vindemiatrix (Fieldwork Findings).

Figure 11 Declination of barrow at Ascott-under-Wychwood.[47]

The Hazletons
Turning to my third site at Hazleton, there were actually two Hazleton barrows, Hazleton North and Hazleton South (Fig. 12). Hazleton North has a declination of +10°, which again widely oriented to within a degree or two to the last of Silva's probable rise points for eclipsing Autumn Full Moons on a minor lunar standstill (Fig. 13).[48] Hazleton South's declination of -21.5°/+23° possibly oriented to the rising southern minor lunar

---

46 Silva, 'Equinoctial Full Moon Models'.

47 Benson, *Excavations*, p. 3.

48 Silva, 'Equinoctial Full Moon Models', Fig. 3, p. 5.

standstill which Ruggles suggests stood at -20° at this time.[49] In terms of fixed stars, the Hazletons oriented to Aldebaran, Denebola, Vindemiatrix, Deneb Adige and Sirius.

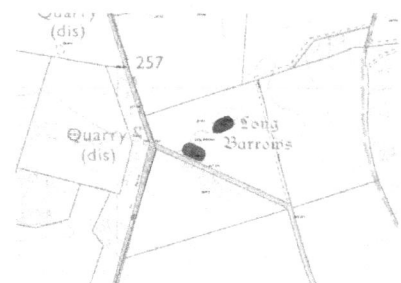

Figure 12 The Hazleton long barrows

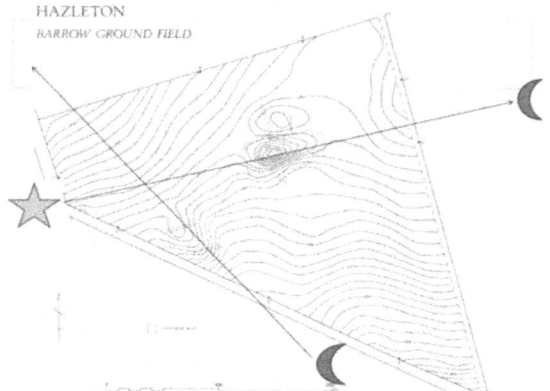

Figure 13 Hazleton North and South's orientations.

Summary of Barrow Findings
Burn Ground's possible orientation to the equinox may indicate a solar astronomy applied at this time, though opinion is divided on equinoctial measurements in general. Ruggles argues that the word equinox should be 'eliminated' from the archaeoastronomer's vocabulary, claiming its use displays a 'highly questionable' tacit assumption that it was meaningful in

---

49 Ruggles, *Prehistoric Astronomy*, p. 57.

prehistoric times.[50] In his view, it is an assumption redolent of Western-style, abstracted conceptions of space and time.[51]

There are other issues to do with probability too. Though my research question focused on lunar and solar celestial events, my findings began to indicate that stellar orientations should perhaps be considered too. As well as orienting to the sun and moon, it became clear that the barrows made just as persuasive a connection to the stars as well. This may be particularly so, given the difficulties of viewing and recording lunar eclipses and minor lunar standstills, especially the second given its nineteen year cycle. The annual rise and set of stars may on the other hand be more dependably observable.

When looking at my findings to this point and given the strictures mentioned above, what can be said is that the barrows orient to the sun, moon and stars, but inferring alignment is problematic. Though Ruggles writes that we cannot hope to understand astronomical practice in pre-historic times without 'beginning to think more seriously' about the people themselves, the information available on the Cotswold landscape is limited. There are no artefacts which indicate that astronomy may have been practiced. My research data represents as much archaeoastronomic evidence as I felt could be justifiably inferred from the sites surveyed.[52]

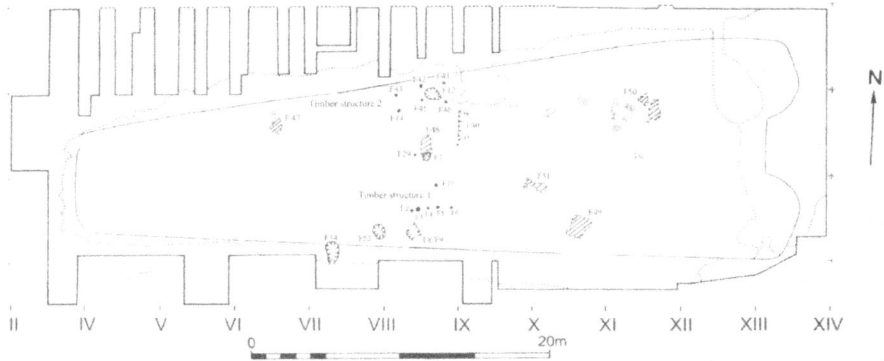

Figure 14 Ascott-under-Wychwood. Numbered post holes in pre barrow context.[53]

---

50 Ruggles, 'Equinox?', p. 45.

51 *Ibid.*, p. 48.

52 Ruggles, *Prehistoric Astronomy*, p. 78.

53 Benson, '*Building and Remembrance*', p. 27.

This brought my survey's primary line of enquiry to a close. However, the archaeological evidence I came across indicated that an exploration of a deeper time profile may prove of interest. The barrows are Neolithic. But as seen earlier at Wayland's Smithy, and as Darvill explains, many long barrows in the Cotswolds and surrounding areas 'seal' earlier structures.'[54] This was the case at Ascott-under-Wychwood, where a complex pattern of Mesolithic holes were found in the pre-barrow context (Fig. 14).

Lesley McFadyen, an archaeologist at Ascott, was struck by the way the Neolithic barrow 'oriented rather uncannily, in the same direction as the Mesolithic post-holes in Timber Structure 1.'[55] Timber Structure 1 is the lowest, single row of post holes (Fig. 15). Unlike Timber Structure 2 they had no adjacent fire pit. This may indicate that this lower row was not domestic, possibly providing a different function such as establishing an orientation to the horizon.

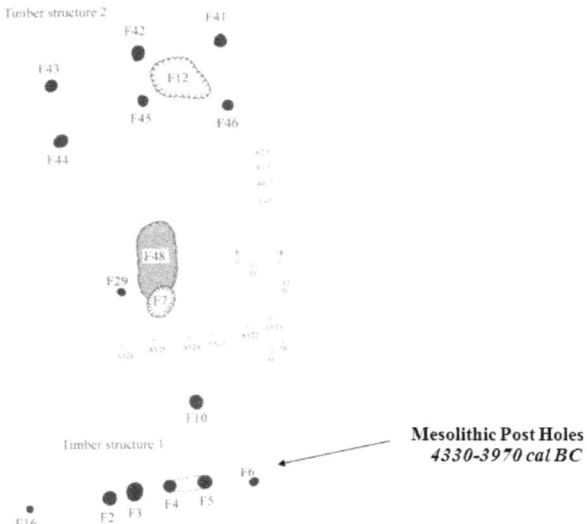

Figure 15 Mesolithic pre-barrow post holes one of which contained beech charcoal, subsequently dated (Benson, *'Building and Remembrance'*, p. 28).

---

54 Darvill, *Cotswolds*, p. 47.

55 Don Benson and Alasdair Whittle Lesley Mcfadyen, 'The Long Barrow', in *Building Memories the Neolithic Cotswold Long Barrow at Ascott-under-Wychwood, Gloucestershire*, ed. Don Benson and Alasdair Whittle (Oxford: Oxbow, 2007), p. 81.

This uncanny replication of orientation noticed by McFadyen occurred between the Mesolithic post holes and the later row of stake holes created by the Neolithic barrow builders, dated 3760–3700 cal BCE (Fig. 16).

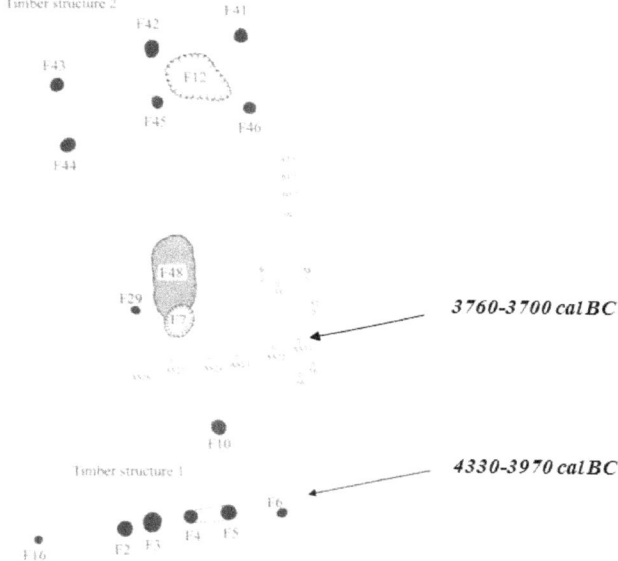

Figure 16 Mesolithic post holes and Neolithic axial divide (Benson, '*Building and Remembrance*', p. 28).

This row of stake holes was the first thing constructed. The line they inscribed on the land established the barrow's fundamental axis, with the final structure subsequently following their angle. This initial marking of the barrow's orientation at its very foundation was typical of the time. North explains the intention behind this process:

> The much flimsier lines of stakes found in so many earthen barrows, clearly mark the stages of construction. The conjecture [being] that the stakes were deliberately set in the directions of lines of sight.[56]

In the case of Ascott-under-Wychwood, as can be seen from the diagram, the orientation of the Neolithic stake holes paralleled the earlier

---

56 North, *Stonehenge*, p. 121.

Mesolithic post holes (Fig. 16). Thus anything up to six hundred years after the Mesolithic orientation was established, the Neolithic monument replicated it. As mentioned, the Ascott-under-Wychwood barrow oriented to the rising Autumn Full Moon, as well as the rising eclipsing Autumn Full Moon at minor lunar standstill, and so did the Mesolithic post holes below it (Fig. 17).

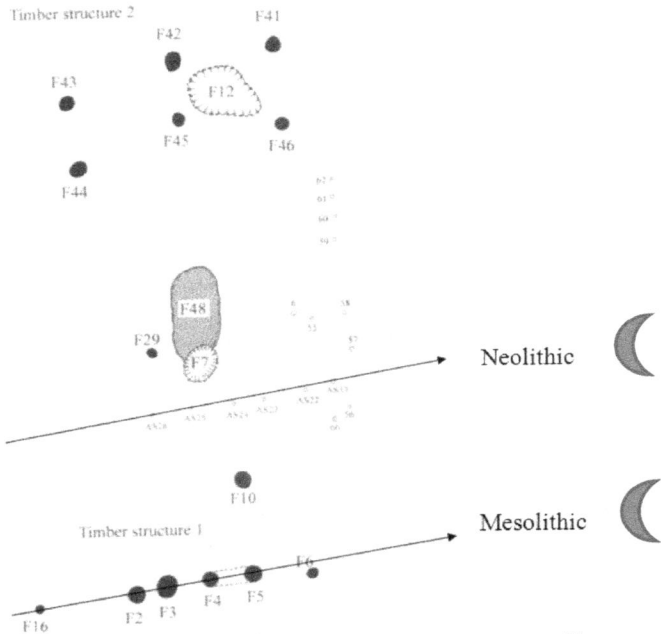

Figure 17 Replication of orientation across eras. [57]

Mesolithic Landscape at Stonehenge

Once this possible astronomic continuity across the Mesolithic/Neolithic transition had been identified, it led to me a second line of enquiry. Though my original research focused on the Neolithic of the north Cotswolds, given that barrows often 'seal' earlier Mesolithic sites I felt that such continuities as clearly applied at Ascott-under-Wychwood may have similarly applied at Stonehenge.[58] The material record at Stonehenge

---

57 Benson, '*Building and Remembrance*', p. 28.

58 Darvill, *Cotswolds*, p. 47.

Pamela Armstrong 47

covers a wide time range and since that record is dated it is possible to establish a similar diachronic profile at Stonehenge as the one I applied to the north Cotswold barrows. The evidence I was looking for would need to be pre-sarsen stone circle, and this is found in the form of three Mesolithic post holes which are currently covered by tarmac in the Stonehenge car park (Fig. 18). The Figure 19 shows the location of posts A / B / C.

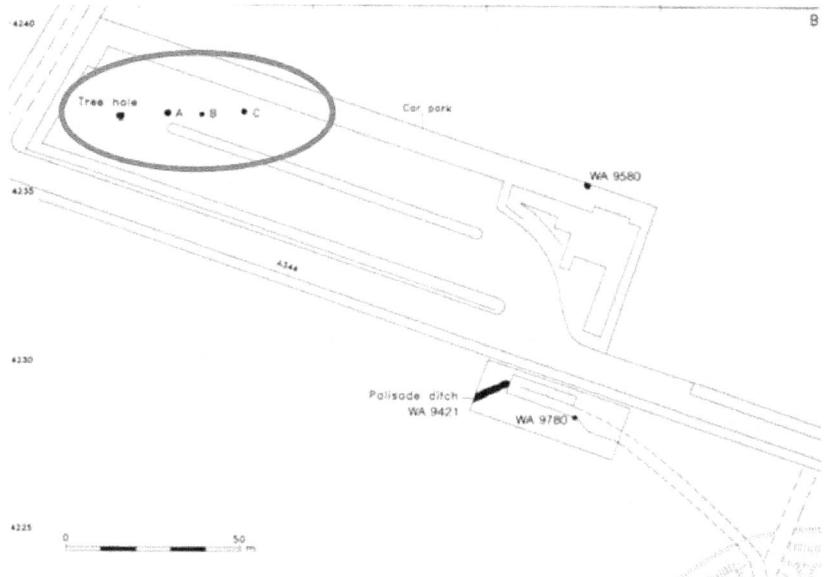

Figure 18 Diagram of car park showing post holes A / B / C.[59]

---

59 Rosamund M. J. Cleal, *Stonhenge in Its Landscape* (London: English Heritage, 1995), p. 42.

48 Skyscapes of the Mesolithic/Neolithic Transition in Western England

Figure 19 Mesolithic car park post holes. Post hole C was undated so was not included in this discussion.

A site survey revealed that the orientation created by posts A and B had a declination of -0°/+1° (Fig. 20).

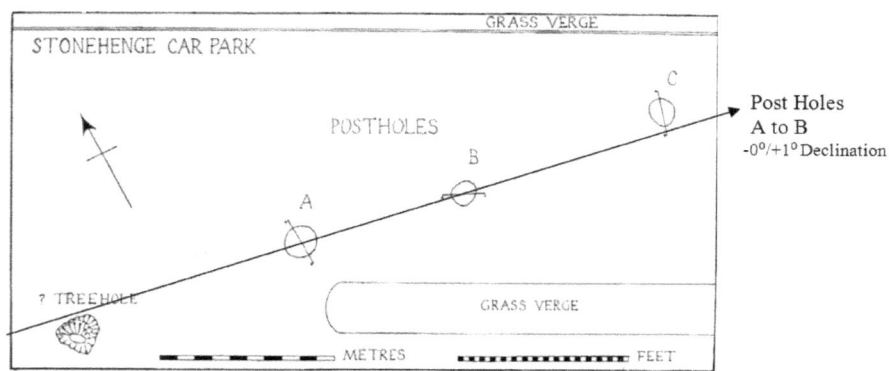

Figure 20 Post Holes A to B. Declination -0°/+1°.[60]

---

60 Roy Loveday, '*The Greater Stonehenge Cursus—the Long View*', *Proceedings of the Prehistoric Society*, Vol. 78, (2012), p. 344.

Loveday's survey of the orientation created by posts A and B also arrived at a measurement of zero degrees of declination. As at Burn Ground, this may indicate an alignment to both lunar and solar equinoctial horizon events.[61] In terms of the fixed stars, Pollux rose at zero degrees of declination at this time.[62] When considering installation of the posts and their sequence, I suggest post A to B is actually the second orientation created at this location. Cleal dates Post B sometime between 7480–6590 cal BCE, but she dates Post A as earlier, sometime between 8820–7730 cal BCE.[63] Thus it is possible that Post A was joined with the tree (Fig. 21) to create the first orientation.

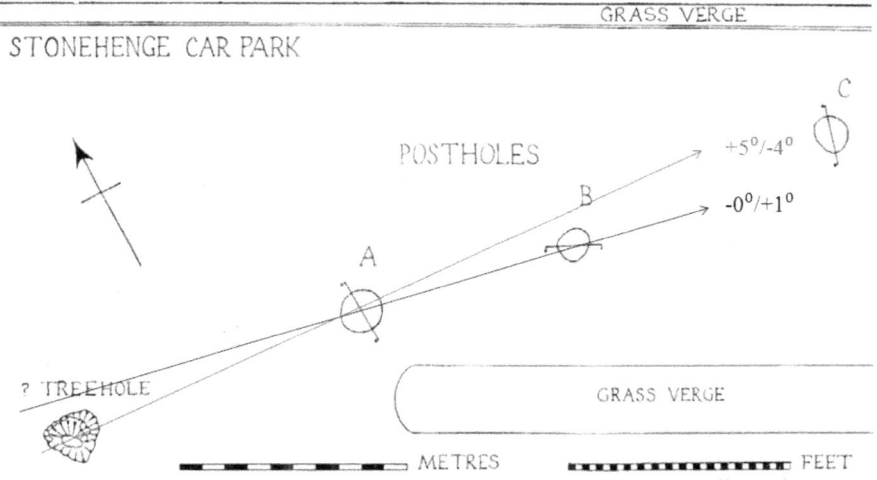

Figure 21 The Tree to Post A's declination is +5°/-4°.[64]

As my fieldwork established a primary azimuth between posts A and B, it was possible to gauge this second orientation of the tree and post A in relation to the first, and that process gave a declination of +5°/-4°. If intended, this was an alignment to the rising Autumn Full Moon and the

---

61 Ibid., p. 345.

62 Stellarium 0.12.0; Silva, 'Equinoctial Full Moon Models'.

63 Cleal, *Stonehenge / Landscape*, p. 43.

64 Loveday, 'The Greater Stonehenge', p. 344.

rising eclipsing Autumn Full Moon at minor lunar standstill.[65] Should this be the case, it suggests that the first orientation established on this Mesolithic hillside was a lunar one, which again supports Sims' theory.[66] However the arrival of a possible equinoctial orientation shown by the second orientation from Post A to B may indicate that a solar astronomy joined the original lunar one. Given the different dates that apply to the post holes this may have occurred without pause or over a period of up to two thousand years, however both orientations were established in the Mesolithic (Fig. 22).

Car Park Alignments

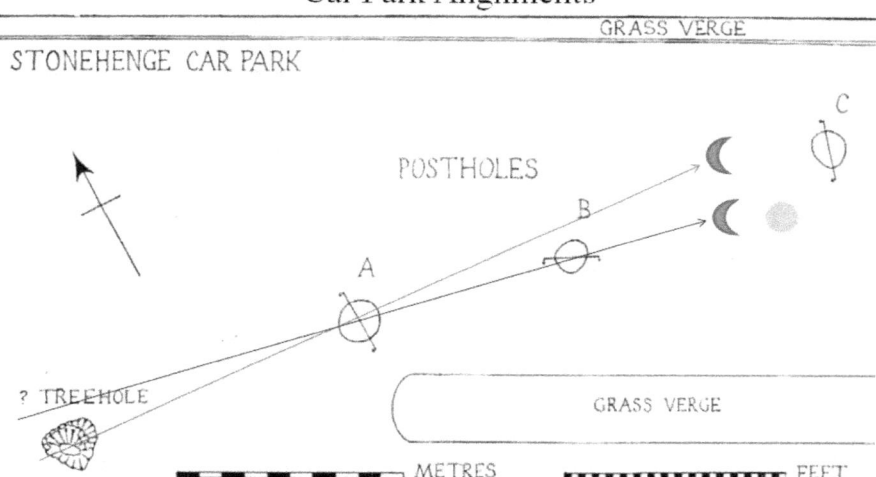

Figure 22 Sun/Moon orientations from the car park. First orientation. Tree to Post A. Lunar. Post A dated 8820–7730 BCE; second orientation. Post A to B. Soli/lunar. Post B dated 7480–6590 BCE.[67]

Summary

In summary, the aim of my survey was to consider the question, 'Does the archaeoastronomic record of the Cotswold-Severn region reflect evidence of a transition from lunar to solar alignment?' Sims' 'solarization' theory

---

65 Silva, 'Equinoctial Full Moon Models'.

66 Sims, 'Solarization', p. 1.

67 Cleal, *Stonehenge / Landscape*. p. 43.

was used as the basis of this study.[68] He argues that in central, southwestern England there was an abrogation from a predominantly lunar to a solar astronomy.[69] Stonehenge was designed, he suggests, in order to engineer this transition.[70] According to Sims, the process of 'solarization' occurred during the sarsen phase of Stonehenge, which dates to around '2413 BCE'.[71] When arguing for this cultural and essentially calendrical shift Sims recommends there be a reinvestigation of evidence further afield than Stonehenge 'for earlier versions of the same complex'.[72] My research has attempted that reinvestigation, focusing on the archaeosastronomic properties of Cotswold-Severn earthen barrows.

The methodology I used to explore these burial chambers was qualitative and hybrid, including fieldwork and in depth analysis of archaeological reports. One of the fundamental aims of my research was to establish a dating sequence in order to contextualise and compare orientations and thus possible alignments. The barrows I chose ranged in date from the very end of the fifth millennium to around 3600 BCE. However, the unexpected emergence of Mesolithic post-hole measurements discovered in pre-barrow contexts prompted me to deepen my time frame. I decided to include a survey of the eighth and ninth millennium post holes in the car park at Stonehenge.[73] If orientation from that earliest time did emerge, it could then be contrasted with evidence from the Neolithic, my latest barrow being dated to '*3710–3655 cal BCE*'.[74] Thus I created a diachronic profile of one small part of the material record across the Mesolithic to Neolithic transition in central southern England. This profile has allowed me to explore the possible astronomies of those who inhabited this region at this time. A number of issues arose during this survey, some to do with the research process itself and others to do with my findings.

---

68 Sims, 'Solarization'. p. 3.

69 Ibid., p. 14.

70 Ibid., p. 3.

71 Cleal, *Stonehenge / Landscape*, p. 231.

72 Sims, 'Solarization', p. 14.

73 Cleal, *Stonehenge / Landscape*, p. 43.

74 Meadows, Barclay, and Bayliss, 'Dating of the Hazleton Long Cairn', p. 54.

52 Skyscapes of the Mesolithic/Neolithic Transition in Western England

The Research Process
Turning first to the hybrid methodology used in this study, it may be of some use to consider its feasibility. The combined use of fieldwork and archaeological report was in direct response to the fragile, ancient material record under investigation. But even though best endeavours were used, it is possible that some calculations are more dependable than others. For instance my measurement for Ascott-under-Wychwood eventuated in an azimuth that at 5°, differed two degrees from that of the report. As Benson, who wrote the report was still alive I both emailed and telephoned him.[75] When questioned about his compass measurement he verbally confirmed and also wrote 'I am confident about the 7 degrees north of east'.[76] I could only source one rough illustration of the Ascott barrow in relation to its nearby road and that was perhaps too blunt a tool to rely on, thus I decided to accept Benson's judgement as final (Fig. 23).

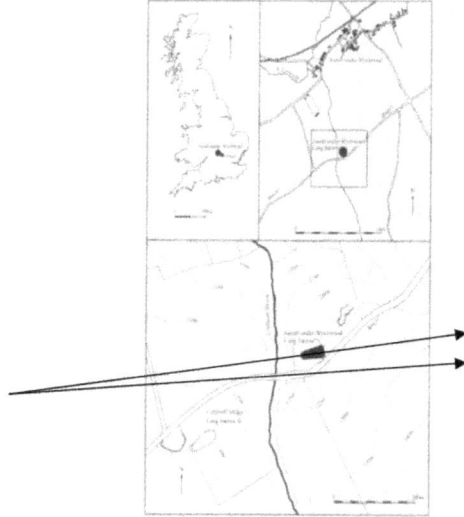

Figure 23 Lower Map, roughly drawn, showing barrow in relation to nearby road.[77]

At Hazleton North on the other hand, the diagrams were finely drawn and I used them with confidence. Combining Savill's illustrations with my

---

75 Don Benson, 11 November 2012.

76 Benson, 4 March 2013.

77 Benson, *Excavations*, p. 3.

own compass measurement of the nearby road I arrived at an azimuth from magnetic north of 79° (Fig. 24).

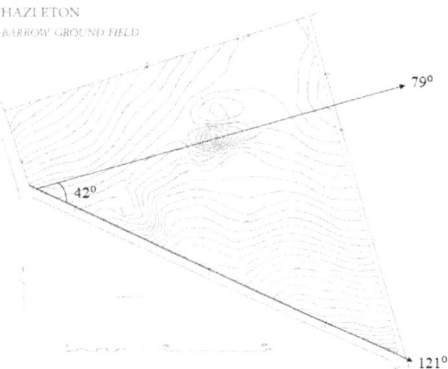

Figure 24 Hazleton North barrow's azimuth of 79° from magnetic north, in relation to adjacent road. Contour survey: contours in metres above OD at 0.25 vertical intervals.[78]

Magnetic North stood at -1.9° on 29th April 2013, the day of my fieldwork so Hazleton North's azimuth, when recalculated, stood at 77°. I subsequently discovered this corresponded with Savill's measurement. He too estimated that the barrow stood at 77°.[79] However, he did not specify which north he was referring to so I emailed him in order to check and had to accept his reply:

> I'm afraid my memory is not up to helping… [One] thing I would add is that the text does say east of north rather than east of magnetic north, but that is perhaps clutching at straws.[80]

So Hazleton North and Ascott-under-Wychwood's measurements were arrived at using the best information a hybrid methodology could source, but it is clear when that methodology's limit has been reached.

---

78 Alan Saville, 'Hazleton North, Gloucestershire, 1979-82 the Excavation of a Neolithic Long Cairn of the Cotswold Severn Group', in *Archaeological Report no 13*, ed. Elizabeth Hall and John Hoyle (English Heritage, 1990), p. 5.

79 Ibid., p. 34.

80 A Saville, 2014, January, Email.

Measurements for the other two sites may perhaps be viewed more confidently. Grimes noted that Burn Ground's 'true axis was almost exactly east-west' and indeed my fieldwork and map usage suggested the barrow had a declination of -0.6°/+0.6°.[81] The same process allowed me to gauge azimuths for the post-holes at Stonehenge. My calculation of 91° corresponded to Loveday's exactly and this led me to more confidently use a diagram to estimate the second orientation at that site.[82] It may be best to assume this level of congruency will not always occur but considering the process overall, hybrid methodologies may find a use with when sensitively applied to appropriate projects.

Research Findings
Returning to the question at the heart of my survey, which asked whether a solar astronomy superseded a lunar one in this region, my findings seem to suggest an attachment to lunar astronomy did apply during the Mesolithic and this continued into the early Neolithic. Sims claims there existed 'an ancient cosmology which in its astronomical aspects had focused on the moon', and indeed, the earliest orientation discovered in my survey was the lunar one on the Mesolithic Stonehenge hillside created by the Tree to Post A.[83] The last lunar orientation was found at my 'youngest' site, Neolithic Hazleton North. This supports Sims theory that a lunar astronomy possibly applied in this region pre-sarsen Stonehenge. However, the emergence of a second orientation on the hillside at Stonehenge, dated to the Mesolithic and this time to zero degrees of declination raises the possibility of solar equinoctial orientations joining lunar ones as early as the eighth millennium BCE. This 'solar' measurement joins with Neolithic Burn Ground, and Hazleton South's equinoctial and solsticial orientations respectively. Sims argues it was 'the Neolithic and Early Bronze Age introduction of solar symbolism *[that]* was to modify and transcend earlier engagement with the moon'.[84] But as the Mesolithic car park post holes

---

81 Grimes, 'Excavations', p. 43.

82 Loveday, 'Greater Stonehenge', p. 345.

83 Sims, 'Solarization', p. 3.

84 Ibid.

Pamela Armstrong 55

clearly predate 'solarizing' mid Neolithic Stonehenge, it is possible that an appreciation of solar horizon event was in place before both those eras.[85]

It is however, necessary to exercise caution when assessing the equinoctial point. As Silva points out, zero degrees of declination can have lunar as well as solar qualities.[86] It is not possible to assume one particular luminary was preferenced at this rise and set point. Given its bimodality, zero degrees remains a declination resistant to definitive interpretation. Separate to the above, another aspect to this declination is that given the sun's speed along this part of the horizon, it is difficult to measure the exact equinoctial point. However, one of my barrows may have been specifically sited to address this problem. Burn Ground was an anomaly in that not only was it the only barrow I surveyed which had zero degrees of declination, it was also the only one located on a completely flat landscape (Figs. 25 and 26).

Figure 25 Burn Ground 180° Panorama. The entire length of the field taken from the south-east. Own photo.

Figure 26 Burn Ground 360° Photographic Panorama. Own photo.

This choice of site was atypical given the generally rolling and hilly Cotswold landscape. It may have been that this particular barrow's zero degree altitude, local horizon was deliberately chosen in order to facilitate the most precise equinoctial measurement possible.

---

85 Cleal, *Stonehenge / Landscape*, p. 43.

86 Silva, 'Equinoctial Full Moon Models'.

One aspect of my research findings which would bear closer study is the use of the word lunar. The 'lunar' orientations which possibly emerge in these findings are to rising Autumn full moon eclipses at Minor Standstill, or a rising Autumn Full Moon. But this phraseology infers an emphasis on the moon, when the sun is just as integral to these events. Full moons and lunar eclipses are simply the culmination of a complex and continuous soli/luni syzygy.

Taking the rising Autumn Full Moon first, Silva argues that combined with the Spring Full Moon, it is one of only two annual celestial horizon events that sees both sun and moon visibly oppose each other across the horizon.[87] As the moon rises at +4° of declination it directly opposes the setting sun at -4° of declination and an 'equinoctial axis' is formed.[88] It is possible that this rare celestial event may have been meaningful in and of itself, but whatever its symbolic function, it involved both luminaries.

Turning to the second 'lunar' orientation which emerged from my study, that was to the rising Autumn Full Moon eclipse at minor standstill. These are eclipses during which the Moon is seen to turn red.[89] They unfold over a number of hours, the actual totality lasting anything up to 72 minutes.[90] As Silva points out, the darkening of a bright Autumn Full Moon at minor standstill is visually arresting.[91] But of note is that these too are equinoctial full moons, they occur just after the sun and the moon are seen to cross over the equinoctial point as they travel in opposite directions along the horizon. C. Marciano Da Silva explains how the horizon relationship between the luminaries is clearly visible at this time. 'One way or the other,' he writes of this full moon, '(it) would be the first full moon past the sun.'[92] Thus as with the non-standstill Autumn Full Moon, both luminaries figure.

---

87 Fabio Silva, 'Equinoctial Full Moons and Solstitial Crescent Moons: An Empircal Luni-Solar Division of the Year?', in *Ninevah to Chaco: Calendars through Time* (Pagosa Springs, 2011).

88 Silva, 'Equinoctial Full Moons and Solstitial Crescent Moons: An Empircal Luni-Solar Division of the Year?', Slide 4.

89 Fabio Silva, 2013, Email.

90 http://blog.nasm.si.edu/astronomy, 'Total Lunar Eclipse'.

91 Silva, 'Equinoctial Full Moon Models', p. 4.

92 Da Silva, 'Spring Full Moon', p. 476.

So on closer inspection it may be inferred that the sun is as implicated in these 'lunar' events as the moon. In both instances the lunar component is indivisible from the solar. Certainly there is no way of establishing which luminary was linguistically prioritised in prehistory. The 'lunar' horizon events possibly revealed in my research happen within days of the autumn equinox, a term currently used to define what is considered a solar calendar moment. But when it comes to describing Equinoctial Full Moons, Silva points out this is, 'the time the sun and moon actually change positions in the sky,' and then he adds, 'In fact, it is possible that EFMs (Equinoctial Full Moons) are the ethnographic definition of equinox'.[93] If this was the case, it suggests an appreciation and experience of sun and moon as indivisible in terms of their relationship each to the other.

Turning to another aspect of my findings and separate to the luni/solar discussion above, though Sims makes no mention of the stars, unexpected but repeated stellar orientation emerged throughout my survey. These were to the very brightest stars on the east, west and northern horizons and suggest that if astronomy was practised at this time, it contained a stellar component. If this was the case there may have been created what Brady terms a 'cosmic and cultural knot.'[94] This supposes a complex relationship with sun, moon and stellar sky lore for navigational, calendrical and ritual purposes.

Lastly, there may be an ethnographic aspect to my findings, given their seasonal emphasis. There was a predominance of orientations to celestial events which occur around the equinoxes. Setting aside observational and record keeping problems, this may indicate that during the Mesolithic to Neolithic transition, trade and ritual gatherings occurred at these times of year. The specific events include a single orientation to an Autumn Full Moon, which was the earliest one found at ninth millennium Stonehenge and subsequent to that all further orientations were to Autumn Full Moon eclipses at minor standstill. This may possibly indicate an astronomy which evolved over time into one that appreciated both eclipses and longer cycles.

In conclusion, my survey aimed to consider the question, 'Does the archaeoastronomic record of the Cotswold-Severn region reflect evidence of a transition from lunar to solar alignment?' Sims' 'solarization' theory was chosen as the originating research for this study, wherein he argued

---

93 Silva, 'Equinoctial Full Moon Models', p. 5.

94 Brady, 'Star Paths', p. 4.

that in central, south-western England there was an abrogation from a predominantly lunar to a solar astronomy.[95] Stonehenge was, he suggests, designed in order to engineer this transition.[96] This position appears to assume a 'lunar' astronomy prevailed pre sarsen Stonehenge, but given the range of orientations which emerge from my survey to possibly the sun, the moon and the stars, it may be possible that a more varied appreciation of the sky existed in those earliest of times.

---

95 Sims, 'Solarization', pp. 3, 14.

96 *Ibid.*, p. 3.

North as a Sacred Direction? Traces of a Prehistoric North-South Route Across Pembrokeshire

Olwyn Pritchard

Abstract: The King's Quoit dolmen perches precariously half-way up a headland in south Wales. Its location has been an enigma since Victorian times. The monument builders chose, not the spectacular sea views of the south-facing slope, but the apparently more mundane inland vista of the north side, with a near horizon comprising a low ridge on the far side of a small valley, now a beach. Previous research by the author into the astronomical possibilities at this site have revealed horizon indicators for cardinal north, in the form of earth mounds which appear to have marked the lower culminations of Deneb and Vega, as they dipped down to the horizon and rose again during the third and fourth millennium BC. This has led to another discovery, namely, that a still traceable route way of roads and footpaths leads north from this monument across Pembrokeshire, passing close by several Neolithic monuments and settlements as it does so, before reaching a sheltered bay and another, south facing, dolmen on the north coast. The southern end of this possible ancient trackway is located on Carmarthen Bay, and the northern end, on Cardigan Bay. Both bays encompass a lengthy stretch of relatively sheltered coastal and estuarine water.

Introduction

The King's Quoit dolmen at Manorbier in southwest Wales seems strangely placed. Built on a ledge half-way up a steeply sloping headland, its location has puzzled archaeologists since Victorian times. The monument builders chose, not the spectacular sea views and sunny aspect of the south-facing slope, but the apparently more mundane inland vista of the north side, with a near horizon comprising a low ridge on the far side of a small valley, now a beach.

Previous research by the author into the archaeoastronomy of this site revealed possible markers for cardinal north, in the form of a conjunction of the local horizon, as seen from the front of the monument, with the lower culminations of the circumpolar stars Deneb and Vega, as they would have been observed during the third and fourth millennium BC.[1]

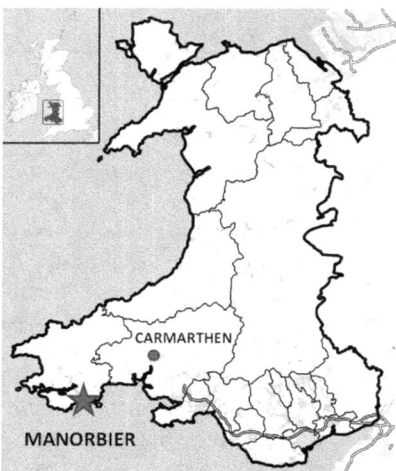

Figure 1 Location map showing position of Manorbier in south Pembrokeshire.

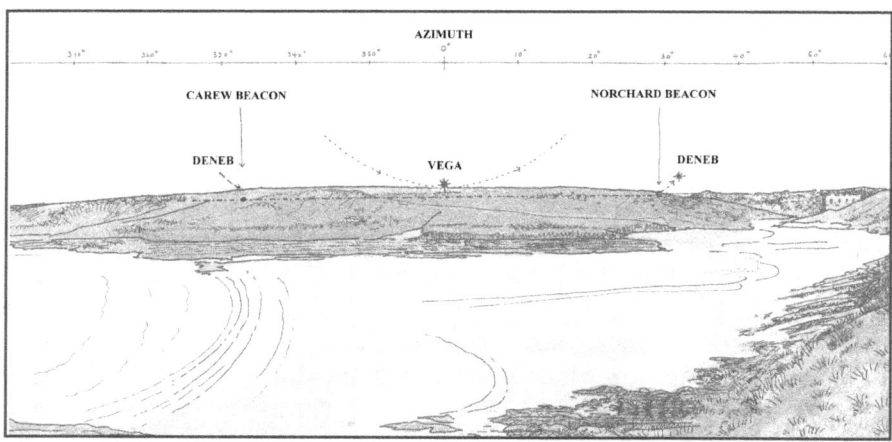

Figure 2 Artists impression of the northern horizon at Manorbier, as viewed from the dolmen, on winter nights during the Neolithic and Early Bronze Age (Illustration by the author, from her own photograph).

---

1 O. Pritchard, 'The Circumpolar Skyscapes of a Pembrokeshire Dolmen', in Fabio Silva and Nicholas Campion, eds. *Skyscapes: The Role and Importance of the Sky in Archaeology* (Oxford: Oxbow Books, 2015), pp. 106–19.

Vega would have just brushed the low ridge on the far side of the valley. Deneb's setting and rising points appear to be marked by two groups of mounds set on an east-west trending ridge of high ground about 3km inland. For an observer standing at the dolmen, this second ridge is just obscured by the nearer ground, but immediately becomes visible to a traveller climbing up the slope on the north side of the beach. From this point, a road leads northwards to the second, inland ridge and looking north again from here, the Preseli hills may be seen on the horizon.

Extending this alignment northwards again, beyond the ridge, and across the peninsula has led to another discovery, which is that a still traceable route way of roads and footpaths leads north from this monument across Pembrokeshire, passing close by several other Neolithic monuments and settlements as it does so, before reaching a sheltered bay and another, south facing, dolmen on the opposite, north facing coast of the peninsula. This second monument, Carreg Coetan Arthur, lies almost due north of the King's Quoit. Manorbier, at the southern end forms a sheltered harbour as does Newport at the northern end on Cardigan Bay.

The Concept of a Transpeninsular Route:
The traditional functional concept of transpeninsular routes is discussed in E. G. Bowen's 1969 book *Saints Seaways and Settlements*.[2] Bowen drew on the earlier work of archaeologists such as Cyril Fox, Lily Chitty, Glyn Daniel, and O. G. S. Crawford.[3] They theorised that the distribution patterns of prehistoric artefacts such as pottery, bronze axes and monuments throughout northwest Europe suggested that people used a network of well-travelled sea routes. Bowen studied these 'trade routes' and showed that they were the same routes later followed in the early

---

2 E. G. Bowen, *Saints Settlements and Seaways in the Celtic Lands* (Cardiff: University of Wales Press, 1969). See also, E. G. Bowen, *Britain and the Western Seaways* (Thames and Hudson, 1972).

3 Cyril Fox, *The Personality of Britain* (Cardiff 1959); Lily Chitty, 'Notes on the Irish Affinities of Three Type 1a Bronze Age Food Vessels found in Wales' *Bulletin of the Board of Celtic Studies* (1938), and 'Irish Bronze Age Axes Assigned to the Guilsfield Hoard, Montgomeryshire', *Archaeologia Cambrensis* (1965); Glyn Daniel, *Megalith Builders of Western Europe* (London: Harper and Collins, 1958); O.G.S Crawford, 'The Distribution of Early Bronze Age Settlements in Britain. *Geographical Journal*, Vol. 40, (1912): pp. 184–203, and 'Western Seaways', in D. Buxton, ed., *Custom is King: Studies in Honour of R. R. Marett* (1936), pp. 181–200.

medieval by the wandering Christians who became known as saints, which suggests that at this time, the routes were as much spiritual pathways as trade routes.[4] Cunliffe also developed and expanded these theories in 'Facing the Ocean'.[5]

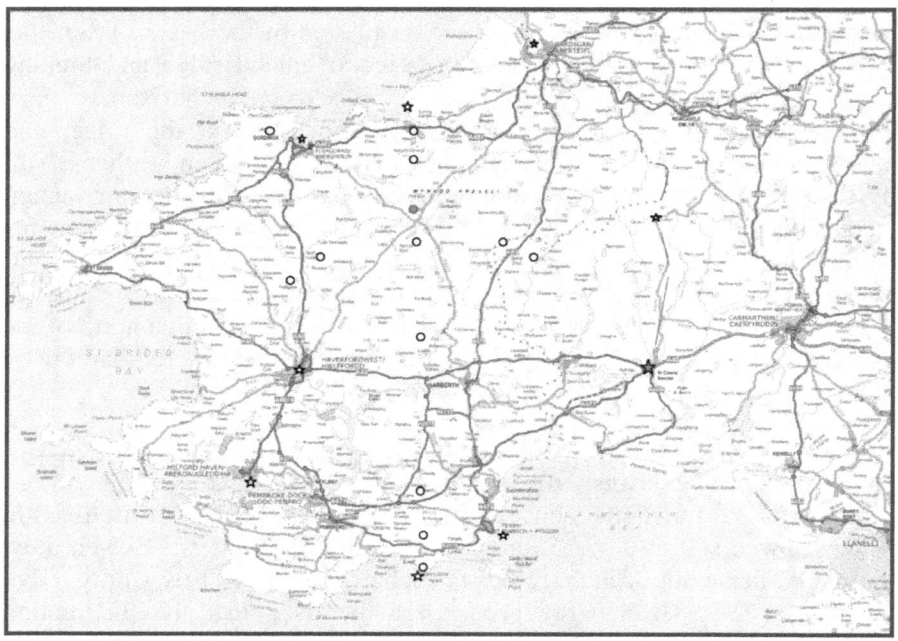

Figure 3 Map of Pembrokeshire and part of Carmarthenshire, showing possible trans peninsular routes of early origin, and their accompanying monuments/ monument complexes, marked by black and yellow circles. The Neolithic causewayed enclosure, Banc Ddu is marked by a red dot, and Bowen's suggested Taf-Teifi route is shown in blue. Stars mark locations referred to in the text. Modern roads in some cases still broadly follow the prehistoric routes. (Edina Digimap).

A particular feature of these routes was the short haul, coast hugging nature of the seagoing sections, incorporating stretches of overland travel, which avoided rounding the ends of peninsulas. The sea conditions off rocky promontories such as Lands End and the Welsh peninsulas are

---

4 Bowen, *Saints Settlements and Seaways in the Celtic Lands*.

5 B. Cunliffe, *Facing the Ocean* (Oxford: Oxford University Press, 2001).

notoriously difficult, prone to stormy weather, heavy seas and strong currents. Sheltered coastal waters and inlets such as Carmarthen Bay or Cardigan Bay are easier for small boats. On the other hand, the inner waters of the bays lack a strong tidal current. Tidal currents have been suggested as an important aid to progress for early human powered craft, so embarking and disembarking from a point towards the end of the peninsula was advantageous from this point of view.[6] Bowen proposed the existence of a transpeninsular route way across southwest Wales, suggesting that it involved the lower reaches of the Rivers Taf and Teifi, linked by an overland section, via the modern settlements of St Clears, Trelech and Cardigan (see Fig. 3).[7]

I would like to suggest that there was at least one other, very early route running almost due north-south across what is now Pembrokeshire between Manorbier and Newport, and that the arrangement of the monuments at Manorbier, which emphasised the circumpolar northern stars acted as an astronomically derived sign post.

Still extant roads and paths follow a course almost due north from Manorbier, although not everywhere in a single line—this course deviates, respecting the landforms, and in some places several parallel track ways form a corridor as much as a mile wide, studded with an unusually dense concentration of monuments chronologically spanning the early Neolithic to the Roman period, including a large Neolithic causewayed enclosure, a rarity in Wales.

Parallels Elsewhere

A similar phenomenon has been noted by Gordon Noble in Scotland, where many early monuments are grouped along the upper Clyde valley, which creates an easy route through the southern highlands.[8] Noble also found that land bridges across two of western Scotland's peninsulas were accompanied by complexes of Neolithic monuments (at Dunragit and Kilmartin). Sherrat suggests that the Wessex monument complex which

---

6 Ibid., pp. 36–38.

7 Bowen, *Saints Settlements and Seaways in the Celtic Lands*, pp. 79–82.

8 Gordon Noble, 'Monumental Journeys: Neolithic Monument Complexes and Route Ways across Scotland', in Vicky Cummings and Robert Johnston, eds., *Prehistoric Journeys* (Oxford: Oxbow, 2007).

incorporates Stonehenge and Avebury can likewise be considered to be situated on a transpeninsular route.[9] This area of Wiltshire lies between the headwaters of three rivers (all called Avon). Traversing these waterways would allow travellers to cross from the English Channel to the Severn without rounding Lands End.[10] Neither Noble nor Sherrat discuss astronomical elements relating to these routes.

The proposed Pembrokeshire route likewise appears to pass through or close by a Neolithic monument complex, with the greatest concentration of sites seven to ten miles inland of the north coast, around a third of the way to the southern end. At seven miles inland of the north coast, the north-south route crosses the Preseli Mountains, passing alongside the earthworks of Banc Du, the only Neolithic causewayed enclosure known in west Wales. In striking parallel, Noble notes that a causewayed enclosure, similarly a rarity in Scotland, is located in the upper Clyde valley, on the transpeninsular route he identified there.[11]

To the south of Banc Ddu, about two miles from the enclosure but within sight of it, the route passes through what must have been an area of intense ritual activity on the lower slopes and southern fringe of the Preseli mountains, judging by the large number of dolmens, cairns and circles recorded there in antiquity. Sadly only a few of these remain today. One surviving monument is the semi-ruinous stone circle known as Dyffryn or Garn Ochr, which, during the Neolithic and Early Bronze Age, would have offered a good view on winter nights of Deneb dipping below the close northern horizon for a few hours before rising from behind the summit of Foel Eryr (the Hill of the Eagles), north and slightly to the east of Banc Ddu. An especially large stone in the northeast quadrant of the circle appears to line up with the summit of Eryr. Vega would have dipped down low over the northern horizon and risen again but not set. Both these observations could have been made throughout the third and fourth millennium BC.

---

9 A. Sherrat, 'Why Wessex? The Avon route and river transport in later British prehistory', *Oxford Journal of Archaeology*, Vol. 15 (1996): pp. 211–34.

10 Ibid., p. 225.

11 Noble, 'Monumental Journeys', pp. 65, 67.

Not Just One Route, but a Network ?
The Preseli Mountains are famous as the source of some of the Bluestones now located at Stonehenge. Both the proposed quarry sites identified recently by Mike Parker Pearson and his team lie within easy walking distance of the Manorbier to Newport route.[12] Professor R. J. C. Atkinson in a discussion relating to the transport of bluestones pointed out that an east-west trackway runs along the top of the ridge, close to the presumed source outcrops, meeting a roughly north-south transpeninsular route still used now—namely the main A 478 road from Cardigan to Tenby, at the eastern end of the ridge.[13] The Cardigan to Tenby road, like the Manorbier to Newport route, also passes through a major Neolithic and Bronze Age monument complex, around Glandy Cross (SN14302660), (see Fig. 3).[14]

The similarities between these two roughly parallel routes could suggest a pattern, especially as the road which links Fishguard and Haverfordwest also runs north-south across the peninsula, to the west of the Manorbier/Newport route, and likewise has an area mid way with striking geological features (e.g., Poll Cairn SM95202452) and numerous prehistoric monuments, including arguably the largest dolmen in Wales, Garn Turne (SM97912727, Archwilio, PRN 2409). The Fishguard/Strumble Head area has a cluster of chambered monuments and the impressive stone row, Parc-y-Meirw (SM99833592), which Thom calculated was aligned on the northern setting point of the minor standstill moon, over Mount Leinster in Wicklow, some 90 miles distant.[15]

As well as the north-south routes, there are signs of early east-west trackways. The path along the Preseli ridge continues westwards beyond Banc Ddu as a minor road to a five way crossroads just east of St David's.

---

12 Mike Parker Pearson, *Bluestone quarry project—2012/2013 excavation report*, Presentation at Pembrokeshire Archaeology Day (Pembroke College, Haverfordwest, 23 Nov 2013).

13 R. J. C. Atkinson, *Stonehenge* (London: Hamish Hamilton, 1956), pp. 174–75, 183–86.

14 T. Kirk, G. Williams, A. Caseldine, J. Crowther, I. Darke, T. Darvill, A. David, K. Murphy, P. Ward, and J. Wilkinson, 'Glandy Cross: A Later Prehistoric Monumental Complex in Carmarthenshire, Wales', *Proceedings of the Prehistoric Society*, Vol. 66, (2000), pp. 257–95.

15 A. Thom, *Megalithic Sites in Britain* (1967; repr. Oxford: Clarendon Press, 1971), p. 159; R. Heath, *Bluestone Magic—A Guide to the Prehistoric Monuments of West Wales* (Llandysul, Wales: Gomer Press, 2010), p. 101.

In the other direction it seems to have continued over high ground until it met the upper Usk valley, then followed the lower ground to modern Chepstow.

This east-west route meets and crosses the north-south, Manorbier/Newport way at Banc Ddu, as do two other current transpeninsular roads making this enclosure site a hub at the crossing point of three local transpeninsular routes and the long distance one. The line of the modern A40, which broadly follows the line of a Roman road west of Carmarthen, and the Fishguard/Newcastle Emlyn/Llanllwni route may represent parallels to the old Preseli track.

The Manorbier to Newport Route and the Monuments
In the following part of this paper I shall give a description of the Manorbier to Newport route as travelled from south to north, with details of the monuments which line it. The objection has been raised by Fabio Silva that the appearance of a roughly linear concentration of monuments stretching across the peninsula may simply reflect a freak of survival, or an act of 'cherry picking'; that originally the distribution of dolmens, earthworks and settlements was equally dense across the entire area, or that those which fall on or near the line have been emphasised by me whereas others which do not, have been overlooked.[16] This is a reasonable argument which is hard to refute entirely, as antiquarian reports show that many monuments throughout the region have been destroyed in the last two hundred years. What appears now to be a pattern may simply be the result of an accident of survival.

In defence of the reality of this route way, however, I offer the presence of the rare causewayed enclosure, Banc Ddu, at what appears to have become a major cross road, and the recorded destruction of a number of monuments actually on the north-south line to the south of this structure (Archwilio PRN1317), which would have further enhanced it.[17] There is a definite cluster of Iron Age sites south of this again, unusual in terms of the sheer numbers in a small area, some of which have been excavated,

---

16 Fabio Silva, pers. comm.

17 C. T. Barker, *The Chambered Tombs of South-West Wales*, (Monograph 14), (Oxford: Oxbow, 1992), p. 51.

showing much earlier origins.[18] This area subsequently attracted a Roman fort.[19]

Additionally, there is the presence of two coastal dolmens—Newport and Manorbier—almost exactly north-south of each other above sheltered harbours, where Mesolithic/Neolithic lithic scatters have been found.[20] Both have their long axes aligned in the direction of travel required to reach the other. The presence of two other parallel north-south roads with accompanying monument complexes nearby is also suggestive of a pattern rather than selective vision, and finally, there have been similar routeways identified elsewhere in the UK, for instance in Scotland and Wiltshire.[21] Further afield, the work of Malville et al. on the Nabta Playa megaliths is especially pertinent as it indicates an apparent concern with the direction of north as well as the use of monuments for navigation during the Neolithic of that region.[22]

---

18 G. Williams, 'Recent work on rural settlement in later prehistoric and early historic Dyfed', *The Antiquaries Journal* Vol. 68, (1988): pp. 30–54; G. Williams, and H. Mytum, 'Llawhaden, Dyfed, Excavation of a small group of defended enclosures, 1980-84', *British Archaeological Reports* 275, (Oxford, 1998).

19 J. Meek, *A Roman Fort at Wiston, Pembrokeshire—Report on First Season's Excavations*, Presentation at Pembrokeshire Archaeology Day, (Pembroke College, Haverfordwest, 23 Nov. 2013).

20 For Newport, see: Archwilio, Dyfed Archaeological Trust online database: http://cofiadurcahcymru.org.uk/arch/dyfed/english/dyfed_interface.html & DAT HER; For Manorbier, see: A. L. Leach, 'Stone Implements from soil drifts and chipping floors etc. in south Pembrokeshire', *Archaeologia Cambrensis*, Vol. 68, (1913): pp. 391–432; V. Cummings, 'Myth, Memory and Metaphor: The Significance of Place, Space, and the Landscape in Mesolithic Pembrokeshire', in R. Young, ed., *Mesolithic Lifeways*, Leicester Archaeological Monographs (2000), pp. 81–86; & DAT HER.

21 Sherrat, 'Why Wessex?'; Noble, 'Monumental Journeys.'

22 J. McKim Malville, R. Schild, F. Wendorf, and R. Brenme, 'Astronomy of Nabta Playa', in J. Holbrook et al., eds., *African Cultural Astronomy—Current Archaeoastronomy and Ethnoastronomy Research in Africa* (New York: Springer, 2007), pp. 131–43.

The Monuments: From the Southern End, Proceeding North

The King's Quoit dolmen (SN 06019728), as mentioned at the beginning of this article, is located on a ledge, half way up the north-facing slope of a steep headland, overlooking a sheltered bay. Barker describes its siting as unusual, but refutes suggestions that it is a natural feature, which observation readily confirms it is not.[23] A large roughly rectangular capstone 5.1m x 2.6m, of locally occurring conglomerate, is supported at the outer end on two uprights of the same rock. The inner end rests on the ground at the foot of a line of natural rock slabs which run across the hillside, beside a third support, from which the capstone has slipped. The long axis of the capstone runs north-south. The northern, outer edge or 'front' faces the near horizon at the far side of the bay. This horizon has an altitude angle of 8° as viewed from beside the dolmen, and would have been grazed by the lower culmination of Vega in the early to mid Neolithic. Concurrently Deneb would have set for a few hours behind this slope, and the positions of its setting and rising, as viewed from the dolmen, appear to have been marked by two groups of mounds on a another, parallel ridge, about two miles inland.[24]

These mounds (e.g., SN07170005, Norchard Beacon) could be reached in less than an hour by a traveller moving northwards inland at normal walking speed. The mounds apparently date from the Early Bronze Age in their current form, but may represent a redevelopment of Neolithic structures—one of several excavated in the 1850s contained a dolmen like structure with a capstone of similar size and shape to that of the presumed Neolithic Kings Quoit, but oriented east west (in parallel to the road running along the ridge).[25] Another mound in the same group produced a highly decorated early Food Vessel, similar to Irish ceramics dating from 1900 to 2300 BC.[26]

Continuing north, the next surviving early monument is some five miles inland, across easy rolling terrain. This is another dolmen, known as the Cuckoo Stones (SN0646 0387), standing on the side of a gentle valley,

---

23 Barker, *The Chambered Tombs of South-West Wales*, p. 38.

24 Pritchard, 'The Circumpolar Skyscapes of a Pembrokeshire Dolmen'.

25 J. Deardon, 'Some Remarks on the Opening of Certain Tumuli Near Tenby', *Archaeologia Cambrensis* (1851): pp. 291–94, here, p. 293.

26 A. Brindley, *The dating of Food Vessels and Urns in Ireland* (Galway: National University of Ireland, 2007), p. 75.

above a stream, on an east facing slope, at 25m OD. Barker finds it a confusing structure, having apparently too many stones for one simple chamber.[27] Nevertheless, it may be found listed on the Archwilio database (PRN 2523) as a Neolithic chambered tomb, and there is certainly a chamber, formed by a capstone propped on uprights, with the open end facing SE, roughly in the direction of winter solstice sunrise. Close by, a stretch of road runs north-south for about two miles and from the top of the slope north of the dolmen, the summit of the highest Preseli Mountain, Foel Cwm Cerwyn may be glimpsed on the horizon.

From here, continuing north, there are no surviving monuments for around seven miles. The line of the north-south route appears to be followed by the modern A4075, traversing more rolling arable fields with wooded valleys and distant views of Preseli from the higher points.

The point where the A4075 meets the east-west trending A40, at Canaston Bridge (SN06661516), marks the southern edge of a notable concentration of defended enclosures, the density of which can be appreciated by viewing the DAT Archwilio map of the area, and selecting for Iron Age sites.

The line of the north-south transpeninsular route continues north through or close by Llawhaden, now the site of a ruined stone 12th century castle (SN0728171746), once the seat of the bishops of St David's, built on the site of an Iron Age defended enclosure. Three miles to the west of Llawhaden lies Wiston, with a motte and bailey castle (SN02261817), also placed within what appears to be a prehistoric enclosure. Recently a Roman fort (SN02501864) has been discovered close to Wiston's medieval remains, straddling an extant road running north-south, with the disused but confirmed Roman road from Carmarthen crossing at right angles—i.e., east-west, adjacent to the fort.[28]

Altogether seventeen Iron Age enclosures are grouped around a two-mile stretch of the transpeninsular route line between the A40 and a steep scarp north of Wiston at Drim (SN06681932). Known as the 'Llawhaden Group' a number of these sites have been excavated. The earliest dateable occupation evidence found was mid–Bronze Age (1500–1300BC), from charcoal in pits sealed by pre-rampart soil at three enclosures, Holgan

---

27 Barker, *The Chambered Tombs of South-West Wales*, p. 47.

28 Meek, *A Roman Fort at Wiston*.

(SN07331818), and nearby Pilcornswell and Woodside.[29] There are also several round mounds in hilltop positions within this area, and single standing stones suggestive of Bronze Age occupation, although none of these sites have been excavated. The enclosures all showed evidence of Iron Age occupation, in some cases extending into the Roman period.

It would seem that this area of the Manorbier to Newport route represented a centre of settlement traversed by north-south and east-west roads during the Iron Age and Roman era. It is also situated between the two arms of the River Cleddau, which flows into the drowned valley of Milford Haven, making it accessible from the sea by boat as well as over land. There are no remaining recognisable Neolithic monuments but as stated, there are signs of Bronze Age activity.

Figure 4 The Preseli Mountains on the northern horizon, as seen from the scarp near Drim.

Descending from the scarp, from which the Preseli Mountains may be seen on the horizon, several linked parallel minor roads lead northwards.

---

29 Williams, 'Recent work on rural settlement'; Williams, and Mytum, 'Llawhaden, Dyfed, Excavation'.

Five miles brings the traveller to the area of New Moat, Henry's Moat, and Bernard's (Brynach's) Well (SN05402800), where the remains of a Neolithic/Bronze age ritual complex sit at the southern edge of the Preseli massif, in an area of powerful and abundant springs. Antiquarian reports tell of multiple dolmens and stone circles in this general area.[30] A dolmen (Holmus Cromlech) was blown up here by a farmer as recently as 1997 (Archwilio, PRN 1317). The Dyffryn circle (SN05882846) survives in a semi ruinous state on the southern side of Bernard's Well Mountain, a low, rounded rock-strewn hill. Climbing to the top gives an excellent view of Banc Ddu (SN06043071), looming up on the northern horizon, although the enclosure is concealed from the circle itself.

Tim Darvill and Geoff Wainwright surveyed Banc Du in 2005 and found the defences irregular, partly interrupted, and not characteristic of the typical Iron Age hill forts of the region. They opened a trial trench across a section of ditch which provided enough material to get Radio Carbon dates from the initial silts which had accumulated in the bottom. These showed that the ditch was open around 3650 BC. Banc Ddu is the first confirmed such Neolithic enclosure in Wales and it is also notable for the fact that visible earthworks still survive above ground level.[31]

To the north of Banc Ddu, the summit of Foel Eryr rises the west side of a pass through the Preseli ridge, at the point where the east-west ridge route mentioned by Atkinson meets the modern B4329 road joining Cardigan and Haverfordwest. The most direct route north to the mouth of the River Nevern, and Carreg Coetan Arthur dolmen, however, passes around the lower slopes of the hill on its western side, following a footpath, which drops down into the upper reaches of the Gwaun valley, crossing the river at Llanerch (SN05703543). The way then ascends the south-facing slope, to pass across Carn Ingli common, and below the notable rocky outcrop of Carn Ingli (SN06293729). To quote Toby Driver of the RCAHMW,

> The stone-built ramparts, enclosures, huts and fields [on Carn Ingli] clearly have their origins in prehistory but, as yet, no excavations have shed light on the development of one of the largest hill forts in west Wales. Three conjoined enclosures on the highest point, are probably the result of multiple periods of early occupation and enlargement. A fourth larger enclosure extends to the north onto

---

30 Done Bushell, 1911; Gardner Wilkinson, 1871.
[31] T. Darvill, G. Wainwright, and T. Driver, 'Among Tombs and Stone Circles on Banc Du', *British Archaeology* (January/February 2007): pp. 26–29.

lower ground and is crowded with stone-built huts and pounds and even the remains of an old street or track. There are twelve gateways, a very high number of vulnerable openings to defend if we assume the structure is an Iron Age hill fort, and it may be that parts of Carn Ingli date back far earlier, to the Neolithic or Bronze Age.[32]

The surrounding area of Carn Ingli Common is also rich in hut circles, cairns and tracks of probable prehistoric origin.[33]

The name Carn Ingli translates as the hill or rock of the angels, becoming associated in the middle ages with St Brynach whose name may be derived from the welsh for hill, 'bryn', and 'iechyd', healing or healthy, and whose church nearby with rock cut cross in the hillside above was an essential way-station for pilgrims en route to St Davids.

Brynach's Well and church at Henry's Moat (SN04412574) to the south of Banc Ddu, are also close to the line of the transpeninsular route.

Below Carn Ingli, where the transpeninsular route meets the Nevern estuary, stands the last (or first) dolmen, Carreg Coetan Arthur (SN06023936), facing slightly east of south, in the direction of midwinter sunrise, which is also the direction taken by the extant minor road from Newport toward Carn Ingli and the south.

Heath notes that the dolmen's location marks the point where the midsummer full moon at the lunar standstill can be seen to emerge from behind Carn Ingli, and roll along the adjoining ridge.[34] Additionally, the author has confirmed by her own measurements taken inside the monument that if originally open as it now stands, it represents a portal for midwinter rising sun/summer solstice setting sun to shine through along its long axis, while the midsummer rising sun/midwinter setting sun would shine through from side to side.

---

32 Toby Driver, *Pembrokeshire Historic Landscapes from the Air* (Aberystwyth: RCAHMW, 2008), pp. 122–24; Coflein – RCAHMW online database http://www.coflein.gov.uk

33 Driver, *Pembrokeshire Historic Landscapes*.

34 Heath, *Bluestone Magic*, p. 105.

Figure 5 Carreg Coetan Arthur dolmen at Newport Pembrokeshire.

The area around the dolmen was excavated by Sian Rees in 1980 and no evidence of a covering mound was seen. Charcoal from secure contexts beneath the floor of the structure gave dates of around 3500 BC. This site also produced sherds from four round bottomed bowls, considered to be broadly contemporary, similar to bowls found at other securely dated early Neolithic sites in the region.[35]

From here it is a few yards only to the tidal waterway leading out into Cardigan Bay, while the shores of the estuary itself have yielded numerous pieces of Mesolithic/Neolithic worked flint.[36]

Discussion – the significance of the Manorbier to Newport route.
The modern road network presumably follows much older routes for most of its length, and we can see that there are a number of ways to cross this western peninsula. The Manorbier to Newport route is only one of several,

---

35 S. Rees, *Dyfed*, Cadw Welsh Historic Monuments Series, (HMSO, 1992), pp. 15–16.

36 Archwilio.

and the only one not now followed by a main road for most of its length. The reason for this is possibly that it is not the easiest or shortest, suggesting that its original purpose may have been as much ceremonial and symbolic, as practical.

The concept of prehistoric pilgrimage has been discussed by Harding, Bradley, Edmonds, Thomas, and Tilley, among others.[37] That long distance journeys were being made in Britain during the Neolithic is suggested by the discovery of cattle bones and teeth at Durrington Walls in Wessex dating to around 3000 BC, which, following isotopic analysis, were found to have originated in Wales and/or possibly Scotland.[38] The frequent discovery of Neolithic stone axes far from their point of origin could be explained by trade or exchange, but could also suggest a high degree of mobility among their owners.[39]

The decision to travel in a particular direction may derive from ideas of the sacredness of a certain orientation. For instance, Bradley found that among the Saami of northern Europe, each of the cardinal directions assumed a special significance.[40] The direction of north was associated with the sacred, with men, wild animals and winter. The northern sky contains the highest point of their 'cosmic river' or world tree, which connects the three layers of the Saami cosmos, consisting of sky, earth, and underworld.

---

37 J. Harding, 'Later Neolithic Ceremonial Centres, Ritual and Pilgrimage—The Monument Complex of Thornborough, North Yorkshire', in A. Ritchie, ed., *Neolithic Orkney in its European Context* (Cambridge: MacDonald Institute for Archaeological Research, 2000), pp. 31–46; J. Harding, B. Johnston, and G. Goodrick, 'Neolithic Cosmology and the Monument Complex of Thornborough, North Yorkshire', *Archaeoastronomy*, Vol. 20, (2006): pp. 26–51; R. Bradley, 'Pilgrimage in Prehistoric Britain?', in J. Stopford, ed., *Pilgrimage Explored* (Woodbridge, Suffolk: York Medieval Press, 1999); M. Edmonds, *Ancestral Geographies of the Neolithic* (London: Routledge, 2002); J. Thomas, *Rethinking the Neolithic* (Cambridge: Cambridge University Press, 1991); C. Tilley, *A Phenomenology of Landscape*: *Places, Paths and Monuments* (Oxford: Berg, 1994).

38 Mike Parker Pearson, *Stonehenge* (Simon and Schuster, 2012), pp. 120, 121.

39 Edmonds, *Ancestral Geographies of the Neolithic*; Bradley, 'Pilgrimage in Prehistoric Britain?'; Thomas, *Rethinking the Neolithic*.

40 R. Bradley, *An Archaeology of Natural Places* (London and New York: Routledge, 2000), p. 12.

Sacred places in the landscape where votive offerings might be made were in many cases 'located along the routes by which people travelled at different times of the year', and were often additionally marked out by unusual or striking topography.[41]

Gordon, in his paper on orientation, mentions that in contrast to the now more common (Judaeo/Christian influenced) preference for east as a sacred direction, some other traditions hold north as the most sacred of all directions.[42] The Gnostics or Dualists were one example—according to the Gnostics, light, beneficence, and revelations came toward our terrestrial world from the north. The Mandaeans of Iran and Iraq face north if they want to think deeply, and the Manichees of third-century Persia (now Iran) considered that their Tree of Life stood in the north. The Chinese meanwhile still favour south as the most auspicious direction.[43] That cardinal directionality has had significance in the cosmologies of many different cultures past and present (in fact, according to Gordon, almost all cultures) can additionally be seen by the attention paid to the cardinal alignment of square prehistoric structures in India, such as Mohenjo-Daro and Hanamsagar.[44]

North has its own special place for the Saami, and may have had significance for the early Neolithic (4500–3600 BC) inhabitants of Nabta Playa, in what is now southern Egypt, where a stone circle and a number of stone rows indicate north. Anthropomorphically shaped stone slabs stand in the sand, and as well as creating north-south alignments, themselves face roughly north. Other directions indicated at this spot include winter solstice sunset/summer solstice sunrise, and the rising points of various stars.[45]

There are two possible ways to view the Manorbier to Newport route. There is the more functional interpretation which regards such transpeninsular tracks as a way for travellers and traders to avoid the

---

41 Ibid., p. 6.

42 B. L. Gordon, 'Sacred Directions, Orientation, and the Top of the Map', *History of Religions*, Vol. 10, no. 3, (1971): pp. 211–27.

43 Gordon, 'Sacred Directions', pp. 218, 219.

44 Gordon, 'Sacred Directions', p. 211; S. Kak, 'Visions of the Cosmos: Archaeoastronomy in Ancient India', *Journal of Cosmology*, Vol. 9, (2010): pp. 2063–77.

45 Malville et al., 'Astronomy of Nabta Playa'.

stormy seas off the western headlands, and there is the ritual/ceremonial approach. The latter does not necessarily exclude the former, but in the case of this proposed route there are alternative shorter ways to cross the peninsula. The apparent emphasis on the northern stars, especially Deneb and Vega at Manorbier is reflected in the linear directionality of the subsequent pathway, leading ultimately to another dolmen facing landwards at its northern end. This could suggest that Neolithic, and perhaps even earlier, peoples living in this region, made a deliberate choice at certain times to follow a north-south path, marked out by the polar stars at night and the sun by day, as it seems the Saami, and perhaps those who built the Nabta Playa also did, and that these celestially based journeys were as much ceremonial as functional. Marking the track with monuments would reinforce this sacred focus.

Conclusion
It appears that during prehistory, a north-south routeway, marked initially by Neolithic monuments, and passing through a central monument complex, stretched between a dolmen on the south coast, and another on the north coast of Pembrokeshire. The nature of the monuments and subsequent additions suggest that they provided a focus for gatherings, ritual activity and settlement which continued into the Medieval period.
For the purposes of trade and travel, if transporting goods or carrying boats, this route would not be as easy or as short as going via Carmarthen or St Clears to the Teifi. However the concentration of monuments from all periods suggests that it was by no means the road less travelled. Perhaps it had something of the pilgrimage about it, making its difficulty irrelevant, or even part of its attraction, in the same way that rock outcrops favoured during the Neolithic for quarrying axe material seem often to have been the highest and most inaccessible.[46]

More work needs to be done, but it may be that the very act of moving in the cardinal directions, as suggested by the apparent remains of an east-west and north-south network of early routes in Pembrokeshire, was in some way a significant act. Alternatively at the very least it would allow for simple navigation by landmarks, sun and stars, a system which even children could grasp, in a time before compasses and maps. Or perhaps some combination of the two, a mystery for the reader to ponder.

---

46 Edmonds, *Ancestral Geographies of the Neolithic*.

The Islandscape of the Megalithic Temple Structures of Prehistoric Malta

Tore Lomsdalen

Abstract: The exploration of the Mediterranean seascape goes back to the foragers of the early Holocene period around the ninth millennium BCE. However there is no secure evidence of human settlement in the Maltese Archipelago before the end of the sixth millennium BCE. Approximately one thousand years later, the unique style of megalithic structures that later became known as the Temple Period commenced. This period lasted about another millennium, then suddenly halted for no apparent reason, leaving no further trace than the monuments themselves. However, based on the extant material culture—artefacts, iconography and the orientation and location of the temples—there are indications that the Temple Period society may have participated in cosmology that integrates land, sea, and sky. Using thick description, this paper will look at the extent to which prehistoric Maltese cosmology consisted of land, sea and skyscape—probably the three main components of an Islanders' cosmology.

1. Introduction

The aim of this paper is to examine if, and to what extent, land, sea, and sky were integrated elements of a Maltese prehistoric cosmology. It will theoretically examine three main areas: firstly, how the early Sicilian seafarers could have arrived in Malta, seen in the context of the exploration of the Mediterranean sea basin which goes back to the foragers of the early Holocene period around the ninth millennium BCE;[1] secondly, how land and sea were the two most inevitable components of an islander's cosmology;[2] thirdly, to what extent did skyscapes provide an additional

---

[1] Graeme Barker, *The Agricultural Revolution in Prehistory: Why Did Foragers Become Farmers?* (Oxford: Oxford University Press, 2006), p. 335.

[2] Reuben Grima, 'An Iconography of Insularity: A Cosmological Interpretation of Some Images and Spaces in the Late Neolithic Temples of Malta', *Institute of Archeology*, Vol. 12, (2001): p. 56.

Tore Lomsdalen, 'The Islandscape of the Megalithic Temple Structures of Prehistoric Malta', *Culture and Cosmos*, Vol. 17, no. 2, Autumn/Winter 2013, pp. 77–105.
www.CultureAndCosmos.org

component of the prehistoric Maltese cosmology, as deduced from archaeological remains, archaeoastronomy, and landscape.[3] The methodology will evaluate retrieved artefacts, iconography, depicted symbolism, images, and the architectural space and layout of the Neolithic monuments. Further, it will look into the cardinal orientations and alignments to celestial bodies of the monumental structures within a context of land, sea, and skyscapes (the word 'skyscape' with reference to the introduction to the Skyscapes TAG volume) based on observations, astronomical calculations and a literary review.

There is no secure evidence of human settlement in the Maltese Archipelago before the end of the sixth millennium (5200 BCE) and archaeological findings do indicate they arrived from Sicily.[4] However, the types of boats or systems of navigation used for the approximately 80km sea crossings are unknown. During good weather conditions Mt. Etna (3340m) and the Hyblean highland in the southeast of Sicily are visible from Malta and Gozo. The Maltese Archipelago may also be visible from the same geographical areas of Sicily; however, it is unlikely to be observable at sea level.[5]

The period when megalithic compounds were erected in Malta and Gozo, the two main islands of the Maltese Islandscape, is generally known as the Maltese Temple Period and chronologically lasts from about 4100 BCE to about 2500 BCE.[6] Whether or not the structures were temples or not, is open for discussion, however, most scholars researching the monuments seem to accept them as temples and they probably did function, at least in part, as sacred places for worship.[7]

---

3 Tore Lomsdalen, 'Is There Evidcence of Intentionality of Sky Involvment in the Prehistoric Megalithic Sites of Mnajdra in Malta?' (MA Dissertation, University of Wales Trinity Saint David, 2013).

4 David H. Trump, *Malta: Prehistory and Temples*, ed. Photography Daniel Cilia (Malta: Midsea Books, 2002), p. 24.

5 Frank Ventura, Seatravel, 2 September 2013. Personal communication.

6 David H. Trump, 'Dating Malta's Prehistory', in *Malta before History*, ed. Daniel Cilia (Malta: Miranda Pubishers, 2004), p. 230.

7 Tore Lomsdalen, *Sky and Purpose in Prehistoric Malta: Sun and Moon at the Temples of Mnajdra* (Ceredigion, Wales: Sophia Centre Press, 2014).

2. Early maritime activity in the Mediterranean basin

Mediterranean seafaring before the Neolithic period around the seventh millennium BCE constitutes a controversial issue; nevertheless, more recent and systematic research conducted in various parts of the Mediterranean basin gradually opens up a new understanding of late Palaeolithic and early Holocene sea travel.[8]

The earliest sea voyage ever reported refers to tools found in the Indonesian island of Flores, dated from more than 800,000 years ago; the signs of some sort of seafaring by pre-sapiens hominids imply the crossing of an estimated 20km deep-water strait between Bali and Lombok.[9] In the Mediterranean, Fernando Pimenta suggests that several sea crossings appear in Sicily more than 30,000 years ago and, about 15,000 years later in pre-Neolithic sites in Sardinia and Crete, indications of maritime activity across a sea-gap of about 15–20km.[10] Evidence of Mediterranean seafaring during the Younger Dryas (12,800–11,500 BCE) is found in the small amount of obsidian from Melos at the Franchthi cave in the Argolid, two locations separated by 120km and reachable by a 20–35km sea-gap crossing between islets.[11] The presence of obsidian, both on mainland Greece and Aegean island sites, suggests that these exploits included successful return journeys, possibly even a seafaring route.[12] The site Aetokremnos in Cyprus, an island that has never been linked to the mainland in recent geological time, shows human presence as well as the

---

8 Cyprian Broodbank, 'The Origin and Early Development of Mediterranean Maritime Activity', *Journal of Mediterranean Archaeology*, Vol. 19, no. 2, (2006).

9 Fernando Pimenta, 'Astronomy and Navigation', in *Handbook of Archaeoastronomy and Ethnoastronomy*, ed. C.L.N. Ruggles (New York: Springer Science+Business Media, 2014).

10 S. Chilardi et al., 'Fontana Nuova Di Ragusa (Sicily, Italy): Southernmost Aurignacian Site in Europe', *Antiquity*, Vol. 70, no. 269, (1996).

11 Catherine Perlès, *Les Industries Lithiques Taillés De Franchthi (Argolide, Grècce)* (Indiapolis: Indiana University Press, 1987), pp. 142–45; Pimenta, 'Navigation'.

12 N. Laskaris et al., 'Late Pleistocene/Early Holocene Seafaring in the Aegean: New Obsidian Hydration Dates with the Sims-Ss Method', *Journal of Archaeological Science*, Vol. 38, (2011).

dwarf hippopotamus, dated back to the eleventh millennium BCE.[13] The distribution of prehistoric sites in Cyprus indicates an extensive coastal exploration; according to Cyprian Broodbank, 'the fact of repeated and seasonal activity hints at more than a one-off venue', either crossing from the Anatolian coast (65–69km) or the longer crossing from Levant (about 100km).[14] The sea level at that point in time may have been up to 55m lower than at present, but was rising rapidly.[15] What kinds of sea-going vessels were used for these crossings is not known; however, preserved dugout canoes used by hunter-gatherers a few millennia later in north Africa and temperate Europe may be indicative.[16] According to Helen Farr, even though the sea level was lower, the presence of hunter-gatherers on Cyprus suggests that people had the necessary maritime technology and social organisation to undertake such open sea voyages.[17] In the early Neolithic there is evidence of a maritime pioneering colonisation in western Mediterranean Europe as agricultural areas were formed by groups of seafaring colonists who moved along the coastline.[18]

According to Mark Patton one should distinguish conceptually between 'discovery and colonisation'.[19] In prehistory, humans who lived in the Mediterranean basin may have known of the existence of an island and could have visited it periodically without actually colonising it. However, Patton further maintains that, in many cases, a clear distinction between the two—based purely on the archaeological record—may be problematic.

---

13 Broodbank, 'Maritime', pp. 208–9.

14 Ibid., p. 209.

15 Helen R. Farr, 'Island Colonization and Trade in the Mediterranean', in *The Global Origins and Development of Seafaring*, ed. Atholl Anderson, James Barrett, and Katie Boyle, (Cambridge: McDonald Institute for Archaeological Research, 2010), p. 180.

16 Sean McGrail, *Boats of the World: From the Stone Age to Medival Times* (Oxford: Oxford University Press, 2001), p. 173.

17 Farr, 'Colonization', p. 180.

18 João Zilhão, 'Radiocarbon Evidence for Maritime Pioneer Colonization at the Origins of Farming in West Mediterranean Europe', *PANAS*, Vol. 98, no. 24, (2001).

19 Mark Patton, *Islands in Time: Island Sociogeography and Mediterranean Prehistory* (London: Routledge, 1996), p. 36.

Furthermore, the coastline of the Mediterranean basin has changed significantly from the Pleistocene until today; at the Last Glacial Maximum (LGM)—radiocarbon-dated to 21,000–18,000 BCE—the Mediterranean coastline lay 120–130m lower than today's level, creating corridors and land bridges between regions which now are divided by water.[20] At the end of the LGM, Formentera was joined to Ibiza, Menorca to Mallorca, Sardinia to Corsica, the Maltese Archipelago and Egadi groups to Sicily, Elba to the Italian mainland; many of the north and east Aegean islands formed part of a mainland area.[21] By 9000 BCE the Mediterranean coastline had basically reached its present level. Sea crossings before that epoch imply taking consideration of Palaeolithic geographical land and sea formations, whereas Holocene (end of the Pleistocene period about 11,700 BCE) and Mesolithic (from about 9000 BCE) sea travellers would be confronted with more or less the present coastline conditions.

Regarding possible determining factors for colonising an island, Patton has worked out a ratio between the surface area of the island and the distance from the mainland, calling it T/DR: a Target/Distance Ratio model, assuming that the islands with the highest biographic ranking are the most likely to be colonised first.[22] Patton further combines this with a visibility model of:

> A) Islands directly visible from land, like Sicily—the largest island in the Mediterranean and only about three km distant from mainland Italy (T/DR=56.3)—and Lipari (T/DR=0.7).
> B) Islands which can be reached without sailing out of sight of land. A large number of Mediterranean islands are classified in this category: Crete (T/DR=0.8), Cyprus (T/DR=1.7 from the Anatolian coast), Sardinia (T/DR=1.5) and the Maltese islands (T/DR=0.1).
> C) Islands which cannot be reached without sailing out of sight of land with only two islands listed: Pantelleria (T/DR=0.06) and Lampedusa (T/DR=0.03), both ratios measured from Sicily, however, in nautical miles they are closer to Africa.

---

20 Cyprian Broodbank, *The Making of Middle Sea: A History of the Mediterranean from the Beginning to the Emergence of the Classical World* (London: Thames & Hudson, 2013), p. 90.

21 Patton, *Islands*, p. 36.

22 Ibid., pp. 43–48.

Malta, Gozo, Pantelleria, and Lampedusa are non-typical in the sense that they have a low ranking but were colonised early (sixth millennium cal. BCE—see next sections), while some islands with high ranking first registered human activity only in the fourth and third millennium cal. BCE, like Salimis (T/DR=200) in the Agro-Saronic Island, situated only 0.5km from the mainland.[23] These listings indicate that the prehistoric settlers considered other variables than size and distance to an island for colonisation. The relationship between available natural resources, specifically biological and/or minerals, and an island's capacity to support a human population, were probably determining factors for colonisation after first discovery. Trading potential of goods may have been an additional factor. Competition, wars, and political conflicts may also have been a reason for defeated rivals to emigrate.

According to Patton, archaeological evidence shows that colonisation in the Mediterranean does not suggest a gradual and continuous process, but rather a 'punctuated equilibrium' with phases linked to significant economic developments, trade explosion, and social changes.[24] In principle this may have been the case; however the process of colonisation of an island may also have been gradual and involved several temporary visits before settlement.[25]

3. Colonisation in Prehistoric Italy and its Islandscape.

According to Robert Leighton, Italy (after France) has more direct fossil evidence for *Homo sapiens neanderthalensis* than any other European country, well represented both in Lazio and Calabria where early human presence goes back before 730,000 BP.[26] A skeleton from a *Homo heidelbergensis* more than 500,000 years old was found in Altamura in southeast Italy.[27] However, a continuous human presence in Italy seems to lack sustainable evidence. More dateable evidence from the Italian Epigravettian (late Upper Palaeolithic) phases appears in numerous sites

---

23 Ibid., p. 57.

24 Ibid., p. 62.

25 Farr, 'Colonization', p. 182.

26 Robert Leighton, *Sicily before History: An Archaeological Survey from the Palaeolithic to the Iron Age* (London: Duckworth, 1999), p. 22.

27 Broodbank, *Middle Sea*, p. 96.

and shows evidence of regional variations in hunting and gathering; a characteristic of this period is an increasing preoccupation with cave art, decorated artefacts, and burials.[28] As Broodbank suggests, the Italian Neanderthals vanished long before 30,000 BCE.[29] By the later stages of the Upper Palaeolithic, Sicily was widely inhabited, as numerous sites show.[30]

When investigating the early Maltese Islandscape, it is important to see the archipelago's human presence in the context of the Italian Islandscape–Sicily in particular. Archaeological evidence indicates that the first settlers on Malta in the early sixth millennium BCE came from Sicily and lived in caves like the Pleistocene cave Ghar Dalam (180,000 years old) where Stentinello-type pottery has been found.[31] As mentioned previously, during the LGM Malta was terrestrially linked to Sicily; whether Sicily was directly linked with mainland Italy is unclear, as Broodbank states in his 2013 publication.[32] However, in a 2006 paper Broodbank claims that the 72m deep Messina strait survived, perhaps in a narrow or otherwise compromised form.[33] Corsica and Sardinia were also fused to create 'Corsardinia' with the shortest sea-gap to the Italian mainland of 15km at that time, compared to over 50km today.[34]

Corsardinia and Sicily, the two largest islands in the Mediterranean and closer to the mainland then they are today, are the first islands to produce definite signs of human presence going back to the Upper Palaeolithic (and possibly to the Lower Palaeolithic as well, although evidence for the latter has been challenged).[35] Elba has also produced Aurignacean material but was probably, at that time, a part of mainland Italy.[36] At the southern half

---

28 Leighton, *Sicily*, p. 22.

29 Broodbank, *Middle Sea*, p. 116.

30 Leighton, *Sicily*, pp. 22-23.

31 Nadia Fabri, *Ghar Dalam: The Cave, the Museum and the Garden*, Insight Heritage Guides (Malta: Heritage Books, 2007), p. 10.

32 Broodbank, *Middle Sea*, p. 121.

33 Broodbank, 'Maritime', p. 206.

34 Broodbank, *Middle Sea*, p. 121.

35 Broodbank, 'Maritime', pp. 206–7.

36 Ibid., p. 206.

84 The Islandscape of the Megalithic Temple Structures of Prehistoric Malta

of the Corsardinian block of the Corbeddu cave, a human phalanx sandwiched between stratigraphic levels is dated to 30,000 years ago; further confirmation of human presence comes from a newer find at Santa Maria de is Acquas in southern Sardinia, comprised of upper Palaeolithic chert and flint tools datable to the LGM (or even earlier).[37] The first human presences on Corsardinia appear to have been temporary ventures; definite signs of settlements in Corsica and Sardinia do not appear until the Mesolithic of the late ninth millennium BCE.[38]

Sicily shows a model of human presence similar to Corsardinia, in that early human habitations seem to be of periodic, not permanent settlements. One site on Sicily may be much older than the rest; Fontana Nuova di Ragusa, a small rock-shelter in the south-western part of the island, shows human occupation going back to the Aurignacian phase (31,700 BP to its uppermost level 40,000 BP). Humans appear to have arrived from the adjoining mainland and it is, according to Chillardi et al., the southernmost Aurignacian site in Europe.[39] The site resembles a temporary shelter and the occupants relied heavily for their nutrition on deer (over 90% of the excavated bones).[40] The Gravettian phase (about 25,000 BCE), which succeeded the Aurignacian phase, shows no sign of human occupation in Sicily; it is not until the Final Epigravettian (about 13,000 BCE) or Holocene era that numerous sites can be assigned with archaeological confidence.[41] Pleistocene fauna of dwarf elephants and hippopotami, swans, and dormice of gigantic size were present in Italy, Sicily, and Malta; such animal bones have also been found in the Pleistocene Maltese Ghar Dalam cave.[42]

When it comes to colonising the smaller islands around Sicily and the Aeolian group there seems to be more a lack of motivation than ability. These islands did not provide a substantial stock of game and wild animals, and had little to offer hunter-gatherers; however three Italian islands were an exception: Palmarola (located 10km off the cost of Lazio), Pantelleria,

---

37 Ibid.

38 Ibid., p. 207.

39 Chilardi et al., 'Fontana Nuova', p. 553.

40 Leighton, *Sicily*, p. 24.

41 Ibid., pp. 25–26.

42 Ibid., p. 17; Trump, *Malta*, p. 56.

Culture and Cosmos

and Lipari (visible from north-eastern Sicily at a distance of about 20km), which were all sources for the precious, naturally occurring, volcanic glass—obsidian.[43]

Liparian obsidian dated to the Mesolithic era was found at Perriere Sottano, near Catania on Sicily; however, no settlement is documented before the Early Neolithic.[44] By that time Liparian obsidian was distributed around the Italian peninsula, Malta, across the Adriatic to the Tremiti Islands, Palagruža, Sušac, and to Dalmatia. According to Farr, seafaring can be seen to have been a booming activity as obsidian from Pantelleria circulated to Lampedusa, Malta and Tunisia; obsidian from Palmarola and Sardinia was distributed in the Tyrrhenian and France.[45] Flint came, to a large extent, from Sicily and pottery was mainly of the Stentinello type. Imported obsidian found in Malta seems to be more of the Lipari type; however some specimens also come from Pantelleria.[46] Pantelleria—about 80km off the Tunisian coast and 110km from Sicily—was at least visited, if not settled in the Early Neolithic, as its obsidian is found in Malta, Sicily, and North Africa.[47] More surprising is the Early Neolithic settlement on the much smaller island Lampedusa, 210km from Sicily and 130km from Tunisia, attested to by finds of engraved pottery, resembling Stinetinello ware.[48] Stentinello-type pottery has also been reported in Tunisia.[49]

The Neolithic, apart from important changes in subsistence strategy—going from hunting-gatherering to sedentarism-agriculture—was a time of exploration, trading, and exchanging goods, seafaring and island colonisation; this most likely demanded new designs for boats and rafts which were probably equipped with basic sails.[50] Iconographic and

---

43 Leighton, *Sicily*, p. 28; Helen R. Farr, 'Seafaring as Social Action', *J Mar Arch*, no. 1, (2006): pp. 86–87.

44 Leighton, *Sicily*, p. 33.

45 Farr, 'Colonization', p. 182.

46 Leighton, *Sicily*, p. 73.

47 Ibid.

48 Ibid., p. 74.

49 Ibid.

50 Ibid.

archaeological evidence of prehistoric boats in the central Mediterranean is scarce. The earliest known Italian boat is an oak log boat found on the submerged Neolithic site of La Marmota on the southern side of Lake Bracciano, 35km north of Rome; in addition to log boats, indigenous reed boat-building traditions existed in central Mediterranean areas.[51] The cognitive horizons of Neolithic peoples stretched well beyond their farmsteads; their fight for nutrition and daily survival and was transmuted into new research and exploration of the wider spheres of economic, political, social, and ritual behaviour patterns and activities.

Seafaring is a skill which requires knowledge on a number of different levels.[52] It requires spatial and temporal awareness; cognitive understanding of land, seascape, and the perception of surroundings is vital, especially when traversing open water or when visibility is bad. A land journey can be broken up into phases and days, but in open sea crossings, where there is no island to make a stopover, the seafaring is an immediate, uncompromising, and dynamic venture. Success depends on careful planning, nutrition, the crew's skills and knowledge, and a keen awareness of all the mortally dangerous risks involved. A leader and master navigator, who may have been in charge of the voyage, would need close collaboration with the rest of the crew.

Pimenta lists a number of possible non-instrumental navigator's skill sets, which include, among other things, steering by the stars, keeping course by the Sun, ocean swells, and the wind.[53] Birds' flying patterns, cloud formation, drifting objects, and changes in water coloration could serve as other non-instrumental tools when sailing out of sight of land. The moon, tides, ocean currents, and methods of keeping track of time—calendars and seasonal markers associated with navigation—are related to the solar year and were directly or indirectly derived from the sun's annual motion.[54] Pimenta further maintains that different societies in different parts of the world, on land or at sea, developed different orientation systems, equally successful with or without material maps.[55] Especially in non-literate civilisations, transference of knowledge from one generation to

---

51 Farr, 'Colonization', p. 183.

52 Farr, 'Seafaring', p. 92.

53 Pimenta, 'Navigation'.

54 Ibid.

55 Ibid.

another played a key role in all social organizations, either related to land, sea, or skyscapes.

Obviously in a much later civilisation, the Vikings' transmission of natural sea navigation methods was essential oral knowledge passed down from father to son. The Vikings brought caged birds with them on open sea crossings. Whenever they lost direction or sight of land, they let the birds free as they would instinctively fly to the nearest land or Islandscape. Apparently the Vikings did not use the compass, wind-vanes or any other instruments; however the use of celestial bodies and knowledge of crude astronomic orientation for deep-sea navigation do seem to have been part of their way-finding at sea.[56] However rudimentary, barbarian or illiterate the Vikings may have been, their natural navigational may go back to prehistoric sea crossing periods.

The importance of having conducted pioneering sea voyages, experienced new land and people, and bringing back valuable goods and merchandise with lifes at stake may have given travellers higher social importance and ranking than others in their community. Farr poses the correct question to which we do not know the answer: 'would these people have gained increased status within their village?'[57] This kind of social segmentation and classification does not fit well into the traditional view of a Neolithic social narrative which seems to be based on non-stratified, agricultural groups.[58] Nevertheless, this issue will be elaborated upon in the next section: namely, that the motivating factor behind the construction of the unique Neolithic temples in Malta may actually have been rooted in a chiefdom society with a specific religious and/or sociological aim and driving force.[59]

4. Colonization of the Maltese Islandscape

According to Trump, attempts to establish that an island settlement took place prior to 5000 BCE are pure guesswork, although he maintains that

---

56 George Indruszewski and John godal, 'Maritime Skills and Astronomic Knoledge in the Viking Age Baltic Sea', *Studia Mytholoogica Slavica*, Vol 9, no. 15–39, (2006): pp. 15–16.

57 Farr, 'Colonization', p. 187.

58 Ibid.

59 Colin Renfrew, *Before Civilization: The Radiocarbon Revolution and Prehsitoric Europe* (London: Pimlico, 1973), p. 170.

88 The Islandscape of the Megalithic Temple Structures of Prehistoric Malta

people were sailing and trading in the Mediterranean by 8000 BCE, well before farming was introduced to the area; therefore, earlier colonisation of Malta 'was by no means impossible'.[60] As Reuben Grima says, 'The story of discovery, exploitation and settlement of the small islands (referring to the Maltese Islandscape) by humans is impossible to separate from that of the exploration, and to some extent the mastery, of the seas around them'.[61] Unlike Pantelleria and Lampedusa, which are not visible from any mainland whatsoever, Malta and Gozo are among the most remote islands in the Mediterranean which, while not directly visible from any mainland, on clear days, Sicily's Mount Etna can just about be seen on the horizon from Malta; also, in theory, both islands are inter-visible.[62] The distance between Sicily and Malta is about 50 nautical miles, but in a real sea voyage this could extend to over 70 nautical miles.[63] Observations made from the Sicilian southern coastlines over a longer period of time with optimal atmospheric conditions, together with meteorological circumstances such as orographic cloud formations of the Maltese Archipelago, could have given the early Neolithic observer a considerable amount of knowledge about the possible existence and position of an island.[64]

An elevated point of the Hyblaean hills in the southeast of Sicily, which rises to over 300m within 10km of the sea, has a visibility of more than 50km out to sea and could serve as a detecting area for identifying the Maltese Islandscape. Grima concludes that careful observation in Sicily could, in principle, detect the Maltese Islandscape without putting to sea.[65] The Maltese islands are lost from sight at sea level less than 50km north of Gozo and the Rabat-Dingli uplands, rising to 250m, are theoretically visible up to about 60km away from the sea; however for a navigator crossing from Sicily it would only come into view after the north of Gozo,

---

60 Trump, *Malta*, pp. 23–24.

61 Reuben Grima, 'The Prehistoric Islandscape', in *The Martime History of Malta: The First Millennia*, ed. Charles Cini and Jonathan Borg (Malta: Salesians of Don Bosco and Heritage Malta, 2011), p. 11.

62 Ventura. Personal communication, 2 September 2013.

63 John Cox, 'The Orientations of Prehistoric Temples in Malta and Gozo', *Archaeoastronomy*, Vol. 16, (2001): p. 35.

64 Grima, 'Islandscape', p. 14.

65 Ibid., p. 15.

Culture and Cosmos

indicating Gozo as the most likely navigational way-point, a position it has retained in modern times for seafarers coming from the north.[66]

As Pantelleria and Lampedusa were colonized contemporaneously, if not slightly before Malta, it seems plausible that Neolithic Sicilians also mastered navigational expertise for crossings between Sicily and Malta, rather than depending purely on visibility. Archaeological evidence such as flint, obsidian, and Stentinello-type pottery indicate direct sea crossings between Pantelleria, Lampedusa, Malta and Sicily, implying a network of exchanging goods and trade during the Neolithic.[67]

Some prehistoric peoples' possible navigational skill sets were mentioned previously. In the case of using astronomy for sea crossings from Sicily to Malta, John Cox suggests the first-magnitude star, Fomalhaut, as an attractive candidate; it passed through south at about midnight in the middle of June in the Temple Period and that the summer may be the height of the sailing season (Fig. 1).[68]

Figure 1 Sky map seen from Santa Croce on the southeast shore of Sicily looking south on August 14 at 21:35, 4,500 BCE where Fomalhaut is clearly seen at about 180° south. Map, T. Lomsdalen

---

66 Ibid., p. 16.

67 Leighton, *Sicily*, p. 74; Grima, 'Islandscape', p. 13.

68 Cox, 'Orientations', p. 33.

Figure 2 shows by sailing due south (azimuth 180°) from Santa Croce on the southeast shore of Sicily, the sea vessel will inevitably hit the Maltese Archipelago. Current, wind, and weather conditions off the Sicilian Channel may have created considerable difficulty in keeping a small boat which was either rowed or sailed on a steady course.[69] During the day, to find the cardinal direction of south, the sun at midday may have given the sailors an indication.[70] Mental maps of a sequence of memorised images or a chain of events could be recollected; to be orientated within an external coordinate, it is necessary to create a logical form of spatial knowledge so that perceptual information and images can be matched with it.[71] In the case of navigation this includes not only knowledge of landscape and seascape, currents, prevailing winds, and wave formations, but also of lunar cycles, star courses, and navigational lore to enable speed, drift, and heading to be reckoned. As Farr concludes 'It is no wonder therefore that seafaring has been described as a specialist occupation'.[72]

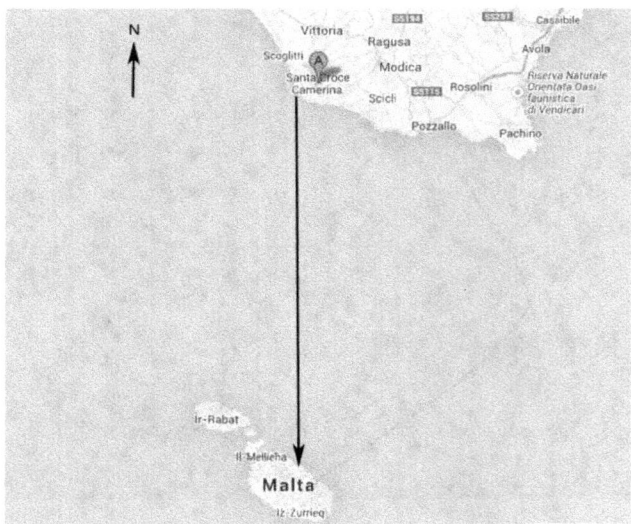

Figure 2 Possible sailing direction from Sicily to Malta.

---

69 Ibid., pp. 33–36.

70 Pimenta, 'Navigation'; Farr, 'Seafaring', p. 92.

71 Farr, 'Seafaring', p. 93.

72 Ibid.

According to Simon Stoddart et al., current archaeological evidence suggests that early Maltese immigrants arrived from the north and colonised a previously unoccupied archipelago; in fact, the colonization of Malta and Gozo could have been seasonal in its occurrence.[73] John Robb suggests that the passage from Sicily to Malta under Neolithic navigational conditions, either rowing or sailing small boats or canoes, would be feasible in one to three days.[74] Robb's suggestion was, in fact, proved valid by Patrick Brydone's trip in 1780; with two companions, three servants, and several hired boatmen, he sailed from Sicily to Malta in a small, oar-propelled boat.[75] At a little after 9:00 PM the boat embarked from Sicily, at about 2:00 AM discovered the island of Malta and, in less than three hours more, reached the city of Valletta. This experimental sea voyage demonstrated that one can row and sail a small oar-propelled craft from Sicily to Malta in less than 24 hours. Unfortunately, Brydone does not mention what kind of navigational system was used for the crossing.

Another experiment in physical inter-visibility between Gozo and Sicily was carried out between July and September 1900 by the Regia Commissione Geodetica Italiana under the direction of the engineer Federico Guarducci.[76] Light signals were sent by means of a lamp from Gozo to the three stations in Sicily; the reply, also by means of light signals, was received from the three stations in Sicily. It is recorded that the signal from Etna, the farthest station, was as bright as a third magnitude star. The calculations gave the distances between the stations and the longitude of Malta was determined from the connection by means of triangulation with Castanea. According to Frank Ventura headlights from cars and streetlamps can be seen on Sicily from Gozo under the right atmospheric conditions.[77] Based on this experiment and observation, the

---

73 Simon Stoddart et al., 'Cult in an Island Society: Prehisrtoric Malta in the Tarxien Period', *Cambridge Archaeological Journal*, Vol. 3, no. 1, (1993): p. 6.

74 John Robb, 'Island Identities: Ritual, Travel and the Creation of Difference in Neolithic Malta', *European Journal of Archaeeology*, Vol. 4, no. 2, (2001): p. 187.

75 Patrick Brydone, *Tour through Sicily and Mata: In a Series of Letters to William Beckford* (1806; repr. London: Fogotten Books, 2012), pp. 177–79.

76 Frank Ventura, *L-Astronomija F'malta* (Malta: Pin, 2002), p. 177. Translated from Maltese to English by Ventura.

77 Ventura, *Visibility*.

theory that Neolithic peoples may have used bonfires as a navigational aid is plausible, if impossible to prove.

The early immigrants would have maintained close contact with Sicily and beyond as certain raw materials, such as obsidian and ochre which do not occur naturally on Malta, are found in the Early Neolithic records.[78] As for Maltese exports, no recognisable objects have yet been identified elsewhere.[79] Robb argues that Malta may have been a sort of trade cul-de-sac, a terminal point in a chain of circulation and re-working of art and ceremonial objects that resulted in a continual importation of primary goods.[80] When it comes to the red ochre which was frequently used in Neolithic burial practices, recent research reveals a high quantity of red ochre on Malta from the Temple Period.[81]

The question of what the Maltese Islandscape could offer the first settlers and why it was selected for immigration will not be extensively discussed in this paper. However, as already mentioned, rivalry, and wars between chiefdom territories in Sicily could be a plausible cause. Exploration of new territories could be another, even if, according to Stoddart et al., Malta and Gozo had little to offer (compared to Sicily) but isolation and poverty of resources.[82] The topographic characteristics would have been no different in prehistory than today; thus, in the period of the first colonisation settlers were presented with an open landscape with little forest.[83] Nevertheless, the island may have been somewhat better covered with soil, offering more abundant natural vegetation than at present.[84] Due

---

78 Robb, 'Identities', p. 187.

79 J. D. Evans, 'Island Archaeology in the Mediterranean: Problems and Opportunities', *World Archaeology*, Vol. 9, no. 1, (1977): p. 20.

80 Robb, 'Identities', p. 188.

81 Nicola A. Montalto et al., 'The Provenancing of Ochres from the Neolithic Temple Period in Malta', *Journal of Archaeological Science*, Vol. 30, no. 1–9, (2012).

82 Stoddart et al., 'Cult', p. 5.

83 Reuben Grima, 'Landscape, Territories, and the Life-Hisotries of Monuments in Temple Period Malta', *Journal of Mediterranean Archeology*, Vol. 21, no. 1, (2008): p. 40.

84 John D. Evans, *The Prehistoric Antiquities of the Maltese Islands: A Survey* (London: The Athlone Press University of London, 1971), p. 3.

to the fragile terrestrial environment and lack of natural resources, this may have prevented colonisation of the Islandscape before the development of a full agriculture and domesticated animal husbandry.[85] If Malta were at all to have received immigration by the Neolithic hunter-gatherers, the transition period to sedentarism and agriculture would most likely have been very short, due to the lack of wildlife for hunting. Domestic animals for breeding were all imported from overseas, most probably Sicily.

5. Temple Period Maltese Cosmology

Today's cosmologists ask the same questions that people have asked for thousands of years.[86] Among those are questions involving the sky. Nicholas Campion links cosmology with astronomy when he suggests that 'the sky is an essential part of human existence. Landscapes do not exist without skyscapes'.[87] Human behaviour may be guided by the belief that life on earth is an imitation of celestial events, and temples are often considered a microcosm of the universe which incarnate and express cosmological beliefs.[88] The recognition of the cyclicality of celestial movements and the deliberate marking of them makes ancient astronomies observable in the ethnographic and archaeological record.

5.1. Land and Seascape

According to Grima the location of the temples in Malta and Gozo, which were often built on south-facing slopes, appears to have been important to their builders.[89] A relationship to the sea seems to prevail with, a marked preference for locations with maritime connectivity, suggesting that the

---

85 Stoddart et al., 'Cult', p. 5.

86 J. McKim Malville, *A Guide to Prehistoric Astronomy in the Southwest* (Boulder, CO: Johnson Books, 2008), p. 3.

87 Nick Campion, 'Locating Archaeoastronomy within Academia', in *Skyscapes: The Role and Importance of the Sky in Archaeology*, eds. F. Silva and N. Campion (Oxford: Oxbow Books, 2015), pp. 59–75.

88 David H. Kelly and Eugene F. Milone, *Exploring Ancient Skies: An Encyclopedic Survey of Archaeoastronomy* (New York: Springer, 2005), p. 2.

89 Reuben Grima, 'Landscape and Ritual in Late Neolithic Malta', in *Cult in Context: Reconsidering Ritual in Archaeology*, eds. David A. Barrowclough and Caroline malone (Oxford: Oxbow Books, 2007), pp. 36–40.

temples might have been a ceremonial gateway between land, sea, and the outside world. This may well have been the framework of an islander's cosmology; Grima defines cosmology as the totality of a belief system.[90]

Stoddart et al. propose the possibility that the Maltese temples were oriented northwest towards Islandscapes of ancestral origins like Sicily, and towards Pantelleria and Lipari as sources of exotic products brought into Malta. However, he does not exclude an alternative interpretation of orientations based on the idea that priests inside the temples may have elaborated a protective and exclusive astronomical lore derived from observations (being southeast) over the shoulders of their congregations who were outside.[91]

Malone maintains that in many cases the temple entrances face a specific direction–usually towards the southeast, south or southwest—and 'that orientation (polarity) clearly makes reference to the celestial world as well as the local topography'.[92]

Grima asserts that 'the specific contexts of an island environment, the ever-present elements of daily experiences are land, sea, and sky', and that travelling in an archipelago environment involves constant interplay with land and sea.[93] Features such as 'valley, river, mountain', as a part of a specific landscape, may be codified into a cosmological scheme. Placing ritual centres such as the prehistoric Maltese temples, with a specific relationship to elements in the topography, positions them in a cosmological scheme of universal significance.[94] Images related to a maritime environment are often located around the courtyards within the temples, while iconography reflecting a terrestrial environment is located within certain temple apses; this suggests a cosmological domain of land and sea, as Grima concludes 'perhaps the two most inevitable components

---

90 Tore Lomsdalen, 'Is There Evidences of Intentionality of Sky Involvment in the Prehistoric Megalithic Sites of Mnajdra in Malta?', Appendix II.

91 Stoddart et al., 'Cult', pp. 16–17.

92 Caroline Malone, 'Metaphor and Maltese Art: Explorations in the Temple Period', *Journal of Mediterranean Archaeology*, Vol. 21, no. 1, (2008): p. 88.

93 Reuben Grima, 'Monuments in Search of a Landscape: The Landscape Context of Monumentality in Late Neolithic Malta' (PhD Thesis, University College London, 2005), p. 247.

94 Grima, 'Landscape', p. 248.

of an islander's cosmology'.[95] Ethnographic evidence suggests that interaction with cosmological representation is not a passive sensory experience of a perceived reality, but an active implementation in the creation of order and meaning in experience and perception.[96]

The figurative representation of Maltese Temple Period art invites interpretations of beliefs, myth, ritual practice, style, and experience. There are different anthropomorphs mainly symbolising human figures, warm-blooded domestic zoomorphs, and monstrous 'other world' creatures, partly real and partly imaginary; some may represent metaphors of life, death, and other worlds, perhaps worlds within worlds. Malone concludes that 'they imply a multi-facetted and many-layered cosmological experience', such as underworlds and cold contrasted with sky/heavens and warmth.[97] Animal representations, wild and domesticated, imaginary or actual, furred, feathered, cold- or warm-blooded, inhabit different layers of a potential cosmos: below the ground, in the sea, on land, and in the sky/heaven, implying an integrated cosmological belief system.[98]

Malta is still a major stopover for seasonally migrating birds.[99] Bird representations, factual or imaginary, are universally identified with flight, especially with spiritual flight and shamanistic trance, taking people through various cosmic levels, making all the cosmos accessible through the art of transformation.[100] Another group represented in Temple Period art are cold-blooded creatures such as fish, snails, lizards, and snakes, which traditionally represent the underworld or the sea.[101] These creatures move between levels of a tiered cosmos and may represent both death and revivification as lizards are associated with sun-seeking and snakes shedding their skins represent a transformation process, mediating between

---

95 Ibid., p. 249.

96 Ibid.

97 Malone, 'Metaphor', p. 92.

98 Ibid., p. 97.

99 Ibid., p. 100.

100 David Lewis-Williams and David Pearce, *Inside the Neolithic Mind: Consciousness, Cosmos and the Real of the Gods* (London: Thames & Hudson, 2005), p. 67.

101 Malone, 'Metaphor', p. 100.

physical and spiritual worlds.[102] Malone suggests that Maltese Temple Period art implies a many-layered cosmos, each layer inhabited by different species/characters.[103] According to Robb, the Maltese temples stood at the conjunction of two systems of cosmological distinctions, 'mediating the above-ground living world and the below-ground ancestral world'.[104] Within this cosmology the distinction between Maltese and 'other' would have occurred through the experience of both temple ritual and overseas travel. The temples may have dominated Maltese cosmological geography and their rise may have involved the construction of a new cosmological value system linked to geographical knowledge and the evolution of a new Islandscape identity based on cosmology.[105] As Mircea Eliade states, stone *is*, always remains itself and strikes man with what is possessed of irreducibility and absoluteness, revealing to man the nature of an absolute existence, beyond time, invulnerable to change.[106]

5.2. Skyscape
The discipline of archaeoastronomy often mingles with the study of prehistory. Clive Ruggles defines archaeoastronomy as 'the study of human perceptions and actions relating to the sky', whereas Kim Malville emphasises that the challenge is to understand the ancient sky watchers and to be able to see the heavens through their eyes.[107] Stanislaw Iwaniszewski suggests 'celestial bodies and phenomena were mentioned in myths and songs, depicted in art, and manipulated as meaningful symbols in rituals and beliefs'.[108]

---

102 Lewis-Williams and Pearce, *Neolithic Mind*, pp. 191–92.

103 Malone, 'Metaphor', p. 105.

104 Robb, 'Identities', p. 191.

105 Ibid., p. 192.

106 Mircea Eliade, *The Sacred and the Profane: The Nature of Religion.* (Orlando: Harcourt, Inc., 1959), pp. 155–56.

107 Clive Ruggles, 'Heavenly Power in Worldly Hands: Ancient Sky Perceptions and Social Control', in *Public Lecture* (Gilching, Germany: European Society for Astronomy in Culture [SEAC], 2010); Malville, *Prehistroic*, p. 3.

108 Stanislaw Iwaniszewski, 'The Sky as a Social Field', in *Archaeoastronomy and Ethnoastronomy: Building Bridges between Cultures*, ed. Clive L. N. Ruggles (Lima, Peru: Cambridge University Press, 2011).

In the previous discussion temple locations were mainly viewed in relation to Islandscape topography and cosmological influences associated with land and seascape. Mario Vassallo, on the other hand, addressed temple locations cosmologically by observing the relationship between sunrise and sunset positions with the topography of the horizon. He concluded that at sixteen out of twenty-four temple sites the winter solstice sun rises at the foot of the first hill to the south of the temple; at five others the sun rises at the point where land and sea meet.[109] Through this study Vassallo implies a universal incorporation of a three-dimensional cosmology into the architectural layout and constructional intentions of the temples, namely land, sea, and skyscapes.

There are substantial indications that prehistoric societies' awareness of astronomical phenomena influenced human behavior. The first hypothesis of a possible relationship between temples and skyscapes came from J. G. Vance, who published his theories in 1842, especially regarding Hagar Qim, but also referring to Mnajdra.[110] Vance suggests that the high north-eastern vertical pillar at Hagar Qim was raised for the purpose of tracing with greater accuracy the motions of different planets.[111] Vance further claims that the temple was never roofed, as the compound was an ideal spot for worshipping the heavenly bodies and paying 'homage to the sun, moon and stars, to dedicate separate temples to each of the two great luminaries, of a like form and contiguous'.[112] By these statements, Vance not only implies an astronomical but also a cosmological connotation to the temples; he further states that decorated slabs next to an altar in Hagar Qim were 'designed to symbolize either the sun or the moon, as being the two great causes of nutrition and generation, or the whole globe of the earth in its widest extent'.[113]

Zammit related the temples to astronomy when, in 1929, he suggested that the pits dug out of a horizontally positioned slab at the entrance to the

---

109 Mario Vassallo, 'The Location of the Maltese Neolithic Temple Sites', *Sunday Times*, 26 August 2007, pp. 44–46.

110 J. G. Vance, 'Description of an Ancient Temple near Crendi, Malta', *Archaeologia*, Vol. 29, (1842): pp. 231–33.

111 Ibid., p. 231.

112 Ibid., pp.232–33.

113 Ibid., pp. 233–34.

98 The Islandscape of the Megalithic Temple Structures of Prehistoric Malta

Tarxien Temple represented an image of the stars of Crux (Southern Cross), a constellation clearly visible from Malta in that period.[114] Luigi Ugolini also indicated, in 1934, a possible relationship between the orientations of the temples and celestial bodies.[115] He also suggests that the Tal-Qadi Stone demonstrates a possible Neolithic 'la astra astrologica', assumingly meaning a piece, sheet, slab, or chart with astronomical or astrological symbolism.[116] From then until 1975 little or nothing happened on the archaeoastronomy front until Gerald Formosa discovered and photographically documented summer solstice sunrise and sunset alignments at Hagar Qim.[117] In the 1980s and 1990s Agius and Ventura analysed possible astronomical alignments of the Maltese temples and measured the central axis orientations of twenty-four temples on Malta and Gozo with a theodolite.[118] Their findings conclude 'it is clear that they are highly non-random', as they were all within less than a quadrant of arc, from Ggantija South, with 125.5° to Mnajdra East with 204°, giving a measure of 78.8° of arc.[119]

In 1990 Paul Micallef published a booklet concluding with clear indications that the Mnajdra South Temple is the only solar temple in the Maltese islands.[120] Richard England suggests that the temple builders' interest in cyclic time through sunrise and sunset not only provided a seasonal timing pattern or marker system to orient the layout and position

---

114 Themistocles Zammit, *The Neolithic Temples of Hal-Tarxien—Malta: A Short Description of the Monuments with Plan and Illustrations*, 3rd ed. (Vallettta: Empire Press, 1929), p. 13.

115 Luigi M. Ugolini, *Malta:Origini Della Civilta Mediterranea* (Malta: La Libreria dello Stato, 1934), p. 128.

116 Ibid., p. 138.

117 Gerald J. Formosa, *The Megalithic Monuments of Malta* (Vancouver, Canada: Skorba, 1975), pp. 17–21.

118 George Agius and Frank Ventura, *Investigation into the Possible Astronomical Aligments of the Copper Age Temples in Malta* (Malta: University Press, 1980).

119 Giorgia Fodera Serio et al., 'The Orientations of the Temples of Malta', *Journal for the History of Astronomy*, Vol. 23, (1992): pp. 116–17.

120 Paul I. Micallef, *Mnajdra Prehistoric Temple: A Calendar in Stone* (Malta: Union Print, 1990), p. 41.

of the temple structures, but also represented celestial archetypes, providing the bridge between mundane time and cosmic time.[121] England further states 'the group ritual force which generated the building forms of Hagar Qim, Mnajdra, and other such sites was born from the belief and conviction that the universe does not function in isolated patterns, but as a whole totally related to the essence of the *cosmos* itself'.[122]

6. Discussion

Aside from the engaging questions of how and why the Neolithic seafarers arrived at an Islandscape, an even more intriguing point of discussion is how they, in the first instance, discovered remote and out-of-sight Islandscapes, such as Pantelleria and Lampedusa. These two islands have the lowest T/D ratio of any Mediterranean island seen from the north—from which direction all archaeological evidence maintains they have been colonised. Their discovery can be compared to finding a needle in a haystack, to use a popular expression.

The scarce research and literature done on sea crossings from Sicily to Malta may give some valuable indications; however, as far as I am aware, little or no research has ever been done regarding Pantelleria and Lampedusa. This might be due to the immense lack of obtainable archaeological, ethno-, and anthropological evidence on the question at hand. The hypothetical criteria prevails too strongly or, to say it pertinently, we have no idea whatsoever! Hopefully, the future will provide us some more indications!

Regarding the hypothetical issues of Lampedusa and Pantelleria, they may have been discovered by pure chance as seafarers lost their way travelling around the central Mediterranean basin, or on their way to/from Europe and Africa. Even so, it is quite impressive that prehistoric peoples placed it on a cognitive map and travelled back to it a second time.

The Neolithic people's possible navigational skill sets have been previously described. In addition, could prehistoric societies have had a cognitive perception of natural elements that moderns lack? Could they challenge and understand basic forces of nature, elements of places in which modern human beings' mental sophistication and Cartesian worldview would not survive? Have we, as a race, lost the ability and

---

121 Richard England, 'A Space-Time Generalogy', in *Malta before History*, ed. Daniel Cilia (Malta: Miranda Publishers, 2004), p. 412.

122 Formosa, *Megalithic*, p. 12.

capability to attune to and live with the natural forces of the world? Is this a plausible explanation for how prehistoric and aboriginal societies explored worlds that most modern human beings would not even dare to consider, nor even survive under the same conditions? Obviously, there is no straight answer to these questions. The same goes for why and how humans crossed the Mediterranean Sea more than 10,000–12,000 years ago.

As a star moves with an apparent speed from east to west at about 10° an hour as seen from earth, it is plausible that sea-navigating star-gazers developed considerable observational knowledge of the rising and setting of stars and star groups over the night sky for certain periods of the year, especially during the major sailing seasons. By following the movements of the sun over a certain period of time, one will notice two periods in a calendar year where days and nights are equally long and the sun rises and sets on exactly the same point; in modern terms these are the spring and autumn equinoxes. These are times of the year when, by using the most basic sundial there is—a *gnomon* or a vertical stick—one can establish the four Cardinal directions: east, west, south and north. This and the stars were used to navigate in the African desert by the nomadic cattle herders at Napta Playa (6100–5500 BCE).[123] Is it plausible that seafarers of that era may have used similar tools for finding their way on the open sea?

When it comes to prehistoric Maltese cosmology there are many indications that the Neolithic temple builders seem to have applied a certain intentionality in the topology of their sacred temple sites represented a multi-level cosmology with land, sea, and skyscapes.[124] There are several indications that Neolithic Maltese communities had, and possibly were inspired by, a cosmic awareness related to the movements of celestial bodies, stars, and star groups, especially the Pleiades with its heliacal rising due east during the Spring.[125] The Pleiades has been

---

123 J. McKim Malville et al., 'Astronomy of Nabta Playa', in *African Cultural Astronomy—Current Archaeoastronomy and Etnoastronomy Research in Africa*, ed. J. Holbrook (Springer, 2008).

124 Grima, 'Landscape', pp. 246–52.

125 Frank Ventura et al., 'Possible Tally Stones at Mnajdra, Malta', *JHA*, Vol. 24, (1993).

universally recognized as having been used by various ancient cultures the world over to mark the passage of time and the seasons of the year.[126]

Based on this theory, it can be argued that taking only land- and seascapes into consideration to represent the totality of Maltese prehistoric cosmology reduces cosmology to a two-dimensional representation of its totality; to say it more directly, to talk about Maltese cosmology without some relation to skyscape causes the term *cosmology* to lose significance as a totality of a belief or universal system, cognitive or spiritual.

Based on the research by Ventura et al. which found, in 1981, an apparent man-made posthole which aligned the Mnajdra South Temple to the winter solstitial sunrise, I took a photo from Mnajdra showing that the sunrise on the winter solstice clearly covers the three cosmological elements—land, sea, and skyscapes—arguably the three main components of an islander's cosmology (Figs. 3 and 4).

Elsewhere I have addressed the hypothesis of astronomical intentionality behind temple constructions and, in particular, the Mnajdra Temple, and concluded: 'The prehistoric temple builders' astronomical purposes or intentionality cannot be verified as there is no written documentation to support such an assumption. Therefore, all evidence is circumstantial, but should not be dismissed merely because it is difficult to quantify'.[127] However, based on research by Ventura et al., Vassallo and others as previously mentioned, there are both quantified and qualified indices that the temples have an orientation and alignments to celestial bodies based on their architecture, constructional layout and location. Further, in my own research on the Mnajdra Temple, only in the South Temple, I found no less than twelve potential alignments towards the rising sun at the Equinoxes and the Solstices.[128] A common and valid argument related to alignments and orientation is; 'it is not difficult to find one if one looks for it'. At Mnajdra, nevertheless, I argue that there are consistent astronomical alignments towards the rising of the sun at specific times of a solar year, throughout its millennia-spanning construction period.[129]

---

126 Anthony Aveni, *People and the Sky: Our Ancestors and the Cosmos* (London: Thames & Hudson, 2008), p. 10; D. R. Dicks, *Early Greek Astronomy to Aristotle* (Ithaca, NY: Cornell University Press, 1970), p. 10.

127 Lomsdalen, 'Intentionality', Ch. 7. Conclusion.

128 Lomsdalen, *Sky and Purpose in Prehistoric Malta*.

129 Ibid.

102 The Islandscape of the Megalithic Temple Structures of Prehistoric Malta

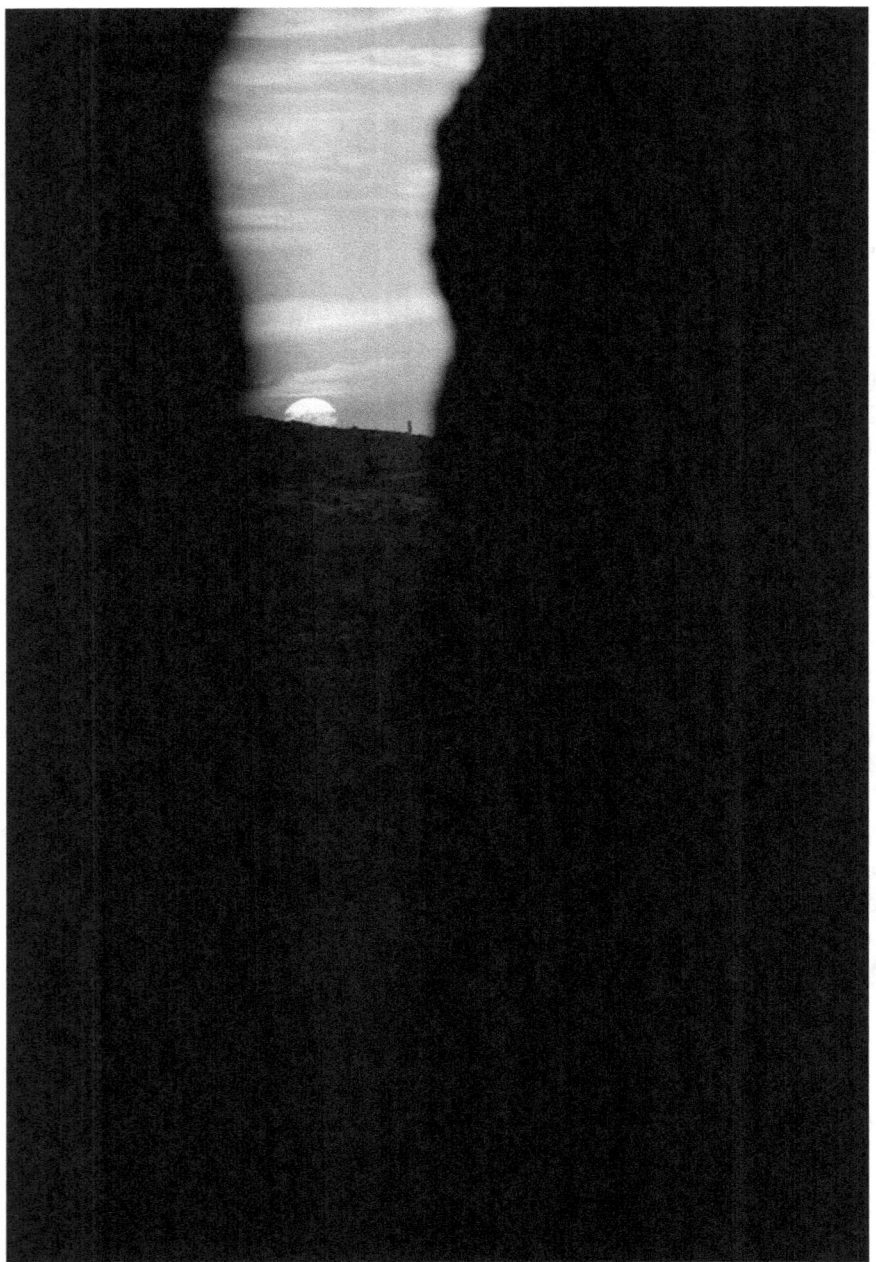

Figure 3 Above the winter solstice sunrise over the posthole seen from inside Mnajdra South Temple. Photo, T. Lomsdalen.

Figure 4 The same winter solstice sunrise [as Fig. 3] seen from inside the Mnajdra Meddle Temple. Photo, T. Lomsdalen.

To include the temple alignments into a cosmological context of the Archipelago's Neolithic population poses and obvious challenge. As Grima has argued earlier, the temples are constructed with demarcated areas of land and sea representations as a part of a cosmological connotation. Malone clearly states they are temples and Stoddart argues they were used for religious rituals with the presence of a priest inside and the rising sun behind the shoulders of the congregation standing outside the temple. Based on physical layout, archaeological artifacts, astronomical observations, circumstantial evidences and various scholarly arguments, the question if the temples were an emic contribution to their builders cosmology, seems to prevail. Whether a temple cosmology was a common religion for the whole Archipelago, is an open question. However, with an estimation of about forty prehistoric temples spread throughout Malta and Gozo, it cannot be disregarded that cosmology may have been an integrated part of the population's belief system and a ritual temple practice.

7. Conclusion

Regarding the objective of this paper to examine if, and to what extent, land, sea and sky were integrated elements of a Maltese prehistoric cultural cosmology, a wide range of relevant scholarly research material and perspectives has been investigated. As outlined, three main areas were taken into consideration to reach a conclusion. Firstly, as Malta is an island and there are many questions regarding how and why this Islandscape was colonised, the wider perspective of early sea travel throughout the Mediterranean basin was discussed; discovery, colonization, and seafaring to and from Malta should be seen in context of the seas around the archipelago. According to material retrieved, the first Mediterranean Islandscape discoveries, visits, and colonisations occurred around 5000 to 6000 years prior to Maltese colonisation. This implies that Mediterranean prehistoric cultures and societies already possessed considerable knowledge and ability of seafaring, long before Maltese settlement. The question of why Malta was colonised at all leaves many more open questions, as the barren Maltese Islandscape seems to have provided few natural resources needed for an increased quality of life for the first settlers.

However, with regard to the Maltese Temple Period's cosmology in context of land, sea, and skyscapes, the picture looks quite different. Malone et al. are probably very close to the truth by stating 'Malta provides one of the best documented cases of prehistoric ritual'.[130] The Maltese Temple Period elaborated a wide range of figurative art, decoration, iconography, and both human and animal representation, together with sacred architecture concerned about both life and death, suggesting a spiritually and cognitively rich worldview, representative of a multi-dimensional cosmology. The temple builders' apparent awareness of celestial bodies' movements and their observations of astronomical phenomena seem to be implemented, represented, and symbolized both ichnographically and in the physical architecture of their temples and ritual structures, manifesting elements of water, earth, heaven, life, and death. This may imply an extended understanding of a spiritual, holistic universe, subject to land, sea, and skyscape: probably, the three most important elements in an islander's cosmology. Nevertheless, further research must be conducted in order to draw definite conclusions on the subject at hand.

---

130 Caroline Malone *et al.*, 'Introduction. Cult in Context', in *Cult in Context: Reconsidering Ritual in Archaeology*, ed. David A. Barrowclough, Caroline Malone (Oxford: Oxbow Books, 2007), p. 3.

Land, Sea and Skyscape: Two Case Studies of Man-made Structures in the Azores Islands

Fernando Pimenta, Nuno Ribeiro, Anabela Joaquinito, António Félix Rodrigues, Antonieta Costa, Fabio Silva

Abstract: The exploration of the Mediterranean seascape goes back to the foragers of the early Holocene period around the ninth millennium BCE. Two case studies in the Azores islands show possible integration of elements of landscape, seascape and skyscape in the way two different types of artificial structures were aligned. The major axes of the Maroiço structures from Pico Island may have been aligned on the summit of Pico Mountain and, reciprocally, on the setting sun at summer solstice over the neighbouring Faial Island. The artificial caves near the sea excavated in Monte Brasil, Terceira Island, may have integrated solar calendrical marks, especially for the Equinox sunset over the distant S. Jorge Island.

Introduction: Human Settlement and Colonisation History

The Azores archipelago is located in the middle of the North Atlantic at a distance of about 1600km from the European continent. The nine main islands are divided into three groups: the Western Group that includes the islands of Flores and Corvo, the Central Group with Terceira, Graciosa, São Jorge, Pico and Faial, and the Eastern group that comprises the islands of São Miguel and Santa Maria (Fig. 1).

Fernando Pimenta, Nuno Ribeiro, Anabela Joaquinito, António Félix Rodrigues, Antonieta Costa, Fabio Silva, 'Land, Sea and Skyscape: Two Case Studies of Man-made Structures in the Azores Islands', *Culture and Cosmos*, Vol. 17, no. 2, Autumn/Winter 2013, pp. 107–32.
www.CultureAndCosmos.org

108 Land, Sea and Skyscape: Two Case Studies of Man-made Structures...

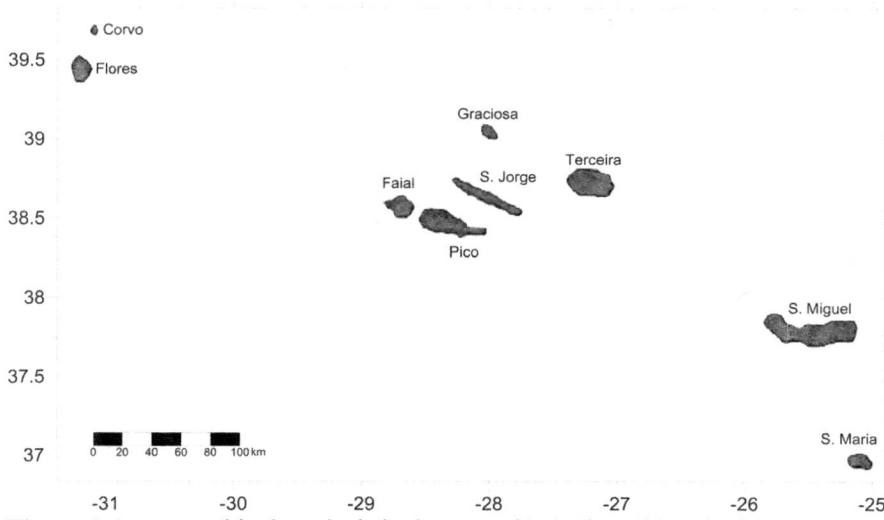

Figure 1 Azores archipelago (axis in degrees of latitude and longitude).

The archipelago is located in a tectonically complex region at the junction of the North American, Eurasian and African plates (Fig.2).

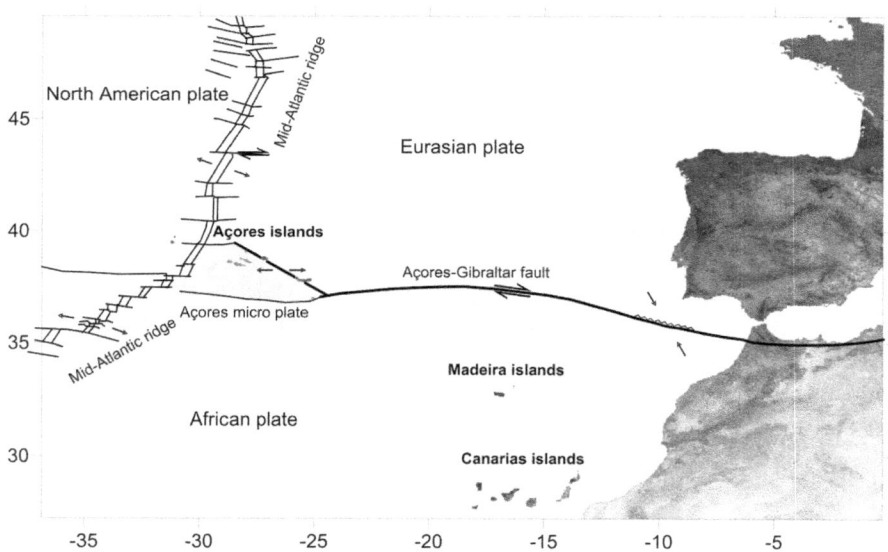

Figure 2 Geotectonic area around Azores archipelago (adapted from Forjaz, 1983 and Buforn et al., 1988; in Nunes, 2000) (axis in degrees of latitude and longitude).

Culture and Cosmos

Portuguese navigators discovered the archipelago in 1432 and colonization began soon after, in 1439. However, the question of whether the Azores were previously inhabited has never been fully explored. References to human presence pre-dating the Portuguese settlement first appeared at the end of sixteenth century. A manuscript by Gaspar Frutuoso refers to a Carthaginian presence but questions whether the Azores were ever part of the mythical archipelago of Atlantis described by Plato.[1] Also at the end of sixteenth century the chronicler Damião de Góis refers to an equestrian statue attributed to the Phoenicians or the Carthaginians in Corvo Island, found when the Portuguese first arrived on the island.[2] References to the Azores, although of questionable veracity, had already appeared in *El Libro del Conoscimiento*, written in the middle of fourteenth century. A set of coins, allegedly found in 1749 in Corvo Island, were bought twelve years later by a numismatist in Madrid, who identified seven coins from Carthage and two from Cyrenaica in an article published in Sweden.[3] More recently, an inscription on what could be the base of a statue is allegedly attributed to the Roman–Dacians.[4] Recent work on the genetic structure of house mice indicates that the majority of M. m. domesticus on Santa Maria and Terceira were found to belong to a D-loop clade indicative of a lineage found in Norway, Iceland, and northern and western parts of the British Isles.[5] All of the above has contributed to the

---

1 Gaspar Frutuoso, *Saudades da Terra*, 6 vols. (Ponta Delgada, Açores: Instituto Cultural de Ponta Delgada, 2005), Vol. 1: pp. 247–80.

2 Damião de Góis, *Crónica do Principe D. João*, ed. Graça Almeida Rodrigues (Lisboa, 1977), Vol. 11: pp. 515–19.

3 J. F. Podolyn, 'Några Anmärkingnar om de Gamles Sjöfart, i anledning af några Carthaginensiska och Cyrenaiska Mynt, fundne år 1749, på en af de Azoriska Öarne', *Det Götheborgska Wetenskaps och Witterhets Samhallets Handlinger Wetenskaps Afdelningen,* (Först Stycket, 1778).

4 Nuno Ribeiro, Anabela Joaquinito, and Sérgio Pereira, 'New Unknown Archaeological Data in Azores: The Hipogeum of the Brazil Mount, Terceira Island (Portugal), and its Parallels with the Cultures of the Mediterranean', Congress SOMA 2012, Florence, Italy. Mediterranean Archaeology. Abstracts; Herbert Sauren, 'Azores: Inscription on the Base of a Statue', 2011 at https://www.yumpu.com/en/document/view/12415814/2011-azores-dacians-netau.

5 Sofia Gabriel, Maria da Luz Mathias, and Jeremy Searle, 'Genetic Structure of House Mouse (Mus musculus Linnaeus 1758) Populations in the Atlantic

debate on whether humans might have settled in the Azores before the Portuguese.

Man-made structures found in some of the islands have for years been attributed to field clearances and other agricultural and pastoral practices of the post-Portuguese colonization periods, despite the complete lack of excavation or study of any surface finds.[6] However, more recently, this take has been challenged by the number, type and characteristics of some of the structures. This paper presents preliminary results from two case studies, particularly focused on their landscape setting and orientation, looking at horizon visibility, axis orientation and possible alignments with landmarks, seamarks and skymarks. These suggest symbolic values that go above and beyond the purely functional roles attributed by previous interpretations.

Case Study 1: The Maroiços of Pico Island
The Portuguese settlement of Pico Island began on its southern side in 1503 and proceeded eastwards and northwards until, in 1723, because of the close trading relationship with the neighbouring island of Faial, the village of Madalena was officially founded adjacent to an important wine production area. Due to poor soils, the production of cereals in the Azores islands was always insufficient for the needs of the populations, especially so in the islands of Pico and S. Jorge. The limited wheat production was complemented by fig, yam and barley cultures and it was not before the middle of seventeenth century that this situation changed when corn was introduced.[7] Corn eventually became the most important crop in the nineteenth century. Vineyards were also planted in the different islands, in volcanic soil not appropriate for wheat or woad (*Isatis tinctoria*) production.

Archipelago of the Azores: Colonization and Dispersal', *Biological Journal of the Linnean Society*, Vol. 108, (2013): pp. 929–40.

6 Nuno Ribeiro, Anabela Joaquinito, Fernando Pimenta, and Romeo Hristov, 'Estudo Histórico Arqueológico sobre as Construções Piramidais existentes no Concelho da Madalena da ilha do Pico (Açores)'. Edição Câmara Municipal da Madalena do Pico (2013), pp. 1–37.

7 Rui de Sousa Martins, 'O pão no Arquipélago dos Açores: mudança e articulação das técnicas de cozedura', *ARQUIPÉLAGO. História*, 2ª série, Vol. 2 (1997): pp. 119–70.

In Pico Island the vineyards, belonging in their majority to owners located in nearby Faial Island, were first planted in the sixteenth century and already produced over two thousand barrels of wine per year, a quantity that rose to an average of twenty thousand by the middle of the seventeenth century. By the end of the eighteenth century it had become the most important wine production area in the Azores Islands: an average of six thousand barrels of Pico wine were exported per year to British America and Antilles.[8] The vineyards were planted in the lava cracks of a coastal strip up to 5km from the sea. The orientation of the vineyards was chosen for the efficient use of sunlight and to profit from the release, during the night, of the energy absorbed by the dark basaltic stones during the day. Rainwater infiltration was maintained by a small layer of soil underneath the lava mantle. After clearing the ground, volcanic stones were used to build walls with a zigzag design to protect the vineyards from the salt-laden winds that, especially from April to June, adversely affected the grape crop. The tremendous effort to build these structures resulted in the Pico vineyard cultural landscape (Fig. 3), now listed as a World Heritage Site.

---

8 Paulo Silveira Sousa, 'Para uma história da vinha e do vinho nos Açores (1750–1950)', *Boletins*, Instituto Histórico da Ilha Terceira, (2004); Carlos Alberto Medeiros, *Finisterra*, Vol. 29, no. 58, (1994): pp. 199–229; Avelino de Freitas de Meneses, 'O vinho na História dos Açores: a introdução, a cultura e a exportação', *ARQUIPÉLAGO. História*, 2ª série, Vol. 14–15 (2010–2011): pp. 177–86; Susana Catarina Silveira Garcia, 'Os alambiques da ilha do Pico, Açores: sistemas técnicos, património e museologia', (Masters Diss., Universidade dos Açores, Ponta Delgada, 2013).

Figure 3 Pico vineyard landscape (with Faial Island seen across the channel).

By the middle of the nineteenth century, after a devastation by Oidium and later by Phylloxera, the vineyards were sold and were replaced by a series of small farms: upper fields were cleared from stones and prepared for corn, yam and potato cultures. According to the local tradition, the monumental pyramidal structures locally called *Maroiços* would date from this period (Fig. 4). The construction of the Maroiços is explained, according to the popular tradition, by the need to clear the land from loose volcanic stones for agricultural use. The smaller stones that could not be arranged in a wall system were piled inside and on the top of large rock piles of pyramidal or conical shapes. These large structures were organized in space together with enclosure walls to limit and protect cultivation areas.[9] According to Garcia, vineyards could be planted on the Maroiços, freeing the terrain for other agricultural products.[10]

---

9 Rui de Sousa Martins, 'Construções de falsa abóbada nas paisagens de pedra seca da Madalena do Pico', *Madalena do Pico—Inventário do Património Imóvel dos Açores*, Jorge Paulus Bruno. coord., (Angra do Heroísmo, Câmara Municipal da Madalena, 2001): pp. 27–30.

10 Garcia, 'Os alambiques da ilha do Pico, Açores'.

Structures similar to the Maroiços can be found in several other volcanic islands like Tenerife, Sicily and the Mauritius Islands. In Tenerife it is accepted, although with some controversy, that the Majanos were built in the middle of nineteenth century, to clear space for cochineal cacti plantations, due to the high market price of this carmine dye.[11] According to Esteban the pyramids in the Mauritius islands may have been built there after the First World War, during a period of prosperity linked to the increase of sugar production.[12]

In other Azorean islands, such as Faial Island, there are structures with construction similar to the Pico Maroiços, but of much smaller size and more irregular shape. Most of the Pico Island Maroiços have staggered steps of about 1 m in height, a lateral ramp and some also have chambered structures inside (Fig. 4). These can divided into the following size categories:

a) Structures about 20 m wide by 6 m tall;
b) Structures about 10 m wide by 13 m tall;
c) Smaller structures between 3–5 m tall.

Figure 4 (left) Maroiço with chambered structure; (middle) Maroiço with staggered steps; (right) complex of Maroiços.

Over a hundred structures are concentrated in an area of six square kilometres, southeast of the town of Madalena (Fig. 5). For each of the 118 locations, the slope and aspect were measured / computed and the azimuth

---

11 César Esteban, 'Arqueología soñada: la historia de las pirámides de Güímar', *el escéptico* (Primavera 2000); Antonio Aparicio and César Esteban, 'Sobre la possible influencia del simbolismo masónico en las orientaciones de las Morras o "Pirámides" de Chacona, en Güímar', *Revista Tabona*, Vol. 17, (enero 2009): pp. 175–87.

12 César Esteban, 'Los Majanos y el don de la ubiquidad', *La Opinión de Tenerife*, Vol. 147, (2002).

profiles of the distance to horizon, horizon height and astronomical declination (corrected for refraction) were generated.

Figure 5 Maroiços cluster to the SE of town of Madalena.

These data show that the structures are located in a terrain with a gentle west-facing slope of +3°, and in locations that present a distant horizon, more than 3 km away, in a range between azimuths 105°–135° and 280°–330° (Fig. 6). From all, the top of Pico Mountain is intervisible to the southeast, at about 120° azimuth.

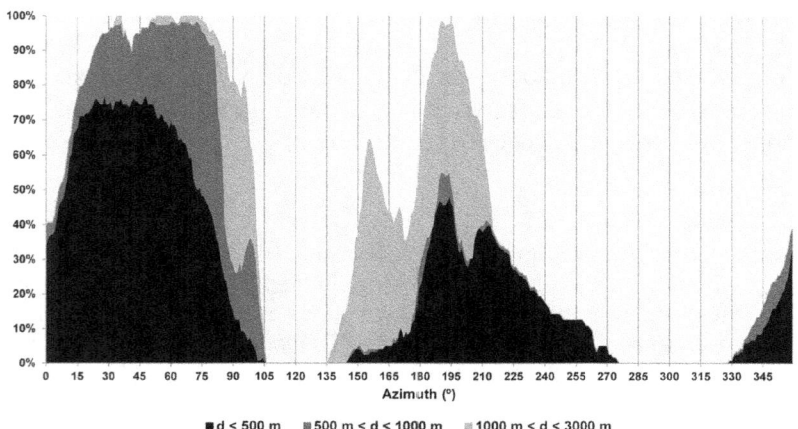

Figure 6 Distance to horizon distribution for the 118 Maroiços in the cluster.

Maroiço orientation

The major axes of the Maroiços were defined by two ranging poles placed at both ends of several steps and at the top of the structure. The azimuth, measured along each pair of poles, was measured with a magnetic compass, and the different measurements were averaged. Magnetic declination correction was estimated using a sub-meter accuracy Differential Global Positioning Sattellite system (DGPS) based on the positions recorded in the Maroiços and a distant topographic feature visible from these Maroiços, to which the magnetic compass bearing was also recorded. The measurements were later compared with an aerial photographic map, orientated to geographical north. The orientation error of the major axis, ranging from 2 to 6 ° in azimuth, was estimated based on the ratio of lengths of major and minor axes, the degree of destruction of the structure and the instrument precision. In some cases the measurements were taken from the aerial photographic map, with a degraded error.

The striking uniformity of the orientations and their possible correlation to prevailing wind directions are illustrated in Figure 7. Prevailing winds in Pico Island are from the southwest quadrant, although from May to July it is common to have slacker winds blowing from the northeast. The strongest winds blow in winter from the southwest or the northwest.[13] The probability of getting a similar distribution from a random circular uniform population is extremely low: p-value of 5E-34 for Chi2 test and 0 for Rao's spacing test.[14]

---

13 João Carlos Carreiro Nunes, 'A Actividade Vulcânica na Ilha do Pico do Plistocénico Superior ao Holocénico: Mecanismo Eruptivo e Hazard Vulcânico', (PhD Thesis, Universidade dos Açores, 2000); Rui Fernando da Costa Medeiros, 'Escoamento do ar em torno da Ilha do Pico e a operacionalidade do seu aeroporto', (Masters Diss., Universidade da Beira Interior, Covilhã, 2009).

14 K. V. Mardia, and P.E. Jupp, 'Statistics of directional data', 2nd ed., (Chicester: John Wiley & Sons, 2000).

116 Land, Sea and Skyscape: Two Case Studies of Man-made Structures...

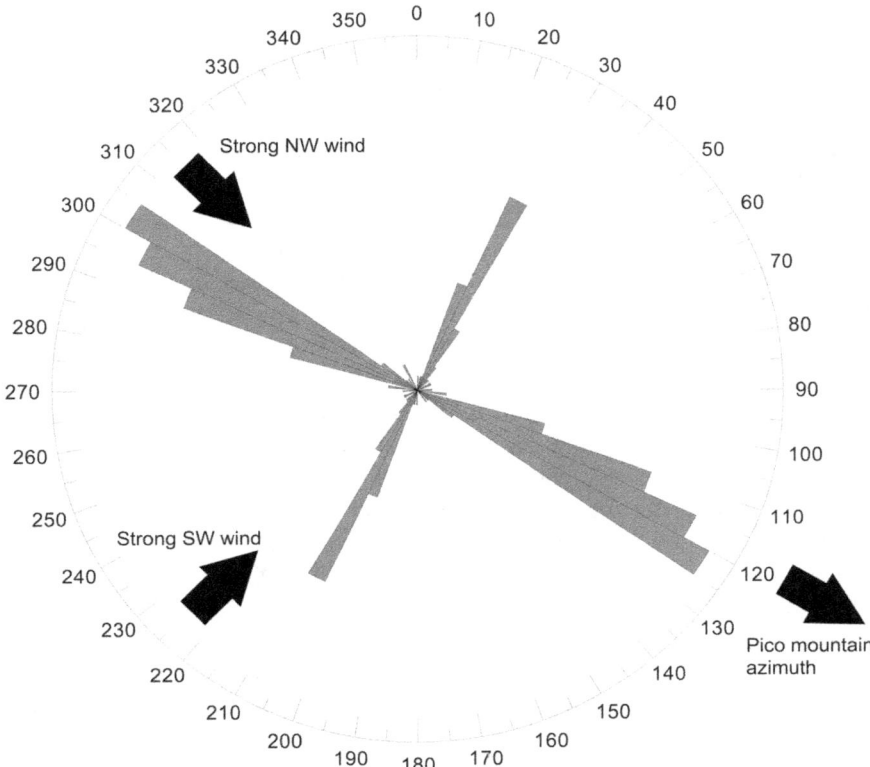

Figure 7 Major axis orientation of the 118 Maroiços in the Madalena cluster. Strong winds occur from NW and SW. Prevailing winds are from SW.

Histograms of orientation were created via a Monte Carlo simulation using the major axis direction of each Maroiço as the mean and the estimated error for each measurement as the standard deviation. These are shown in Fig. 8 (left) for the eastern direction and Fig. 9 (left) for the western direction. In the same figures we represent in red, the azimuth distribution for the Pico mountain-top as seen from each site (Fig. 8) and the azimuth distribution for Summer solstice sunset, calculated with the height of the horizon over Faial Island, as seen from each site (Fig. 9). As the figures make clear the major axes of most Maroiços are oriented in such a way as to align with the Pico mountain top to the east, and summer solstice sunset over Faial Island to the west.

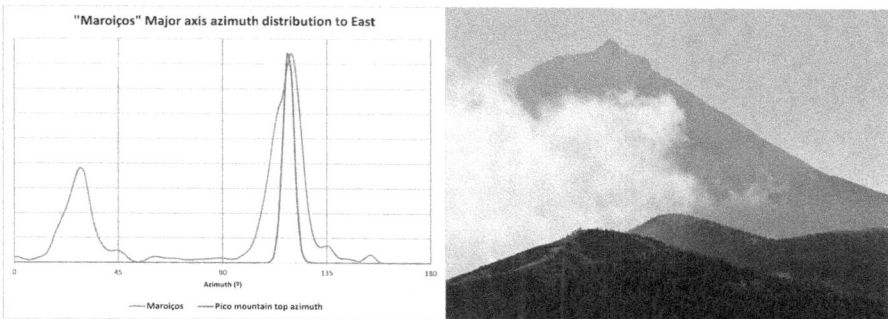

Figure 8 (left) Histogram for the eastern azimuths of the 118 Maroiços in the cluster; (right) Pico Mountain.

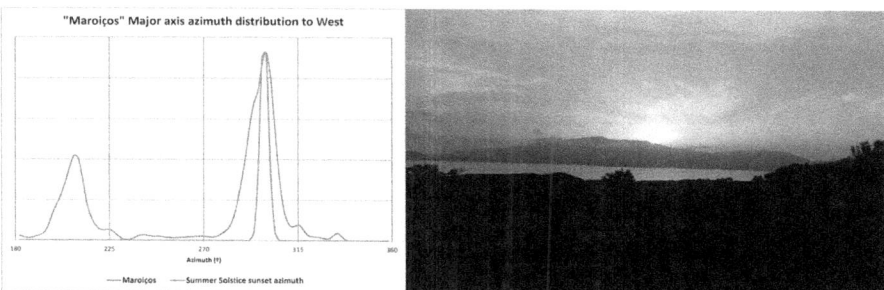

Figure 9 (left) Histogram for the western azimuths of the 118 Maroiços in the cluster; (right) Sunset over Faial Island 11 days after summer solstice 2013.

Discussion

The pattern for the major axis orientations presents strong evidence for an intentional alignment in the construction of the Maroiços. This apparent intentionality may have been driven by ecological factors, such as crop protection from prevailing winds, using a thumb rule for coarse alignment to the mountain top. However we cannot discard a symbolic motivation associated with the summer solstice. The fog, very common in summer, especially in June, blocks the view to the mountain, but this alone would not explain an alignment in the opposite direction to the summer solstice sunset. The fact that the Maroiços were built in locations that present open and reciprocal views to 120° and 300° of azimuth and are concentrated in a region where a dual alignment to the top of Pico mountain and the Summer solstice sunset over the island of Faial is visible, suggests an integration of a possible bidirectional alignment building convention (Figs.10 and 11).

118 Land, Sea and Skyscape: Two Case Studies of Man-made Structures...

Figure 10 Alignment between summer solstice sunset and Pico Mountain peak (yellow line). The Maroiços cluster is marked by a black circle.

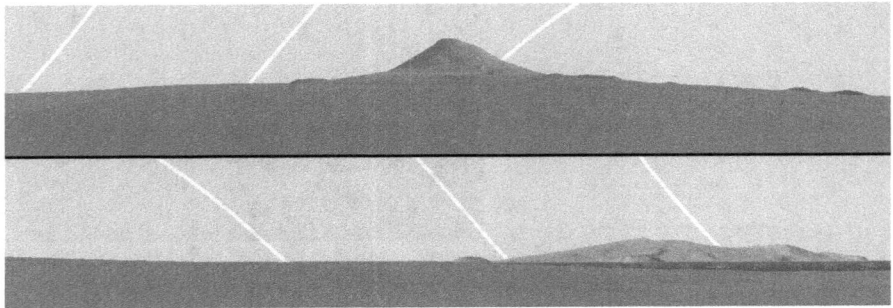

Figure 11 Computer generated panoramas for Maroiço M7 site with sun rise and set orbits marked for the solstices and equinoxes (yellow lines): (top) view towards Pico mountain; (bottom) view towards Faial Island.

An archaeological survey of this cluster during the summer of 2013 provided additional data which is summarized in the appendix.[15] If we accept the radiocarbon dates obtained for structure M7 then it might be difficult to explain why solar alignment symbolism may have been used by such a developed society. The sixteenth- and seventeenth-century dates, although not necessarily synchronous with the construction of the structures, do indicate that at least some of the Maroiços are older than would be expected if those structures were contemporaneous with the foundation of the village of Madalena, or built after the vineyard crisis in

---

15 Ribeiro, et al., 'Estudo Histórico Arqueológico sobre as Construções Piramidais existentes no Concelho da Madalena da ilha do Pico (Açores)'.

the middle of the nineteenth century, when large areas of land were cleared for corn plantations.

There have been associations of the Majanos, in Canaries, to a description written by Juan de Abreu Galindo in 1632 about one Guanche practice in Canaries, where stones were gathered in a pyramidal pile and used in rituals at particular dates in the year.[16] Nevertheless other authors identified these pyramids not with the Majanos but with other structures in Roque de los Muchachos, La Palma, of circular, oval or quadrangular shapes and diameters between 1.5 m and 4 m and up to 1 m height, used by the Guanches for ritual practices and probably related to the piles of stones documented in sites of Gran Canaria and El Hierro, as well as the Berber kerkús.[17]

It is curious that in the Chacona complex, in Güímar, Tenerife, the orientation defined by the common north wall (and also by the major side of the bigger structure) is also in the direction of the summer solstice sunset.[18] The excavations made in the central platform of this complex resulted in a stratigraphy compatible with an agricultural soil, with the total absence of pottery that could be dated before the nineteenth century. The absence of clear written references to the existence of these structures in the Canary Islands before the middle of nineteenth century, led several researchers to associate their construction with the increase of cochineal production in the second half of nineteenth century.[19] In the particular case of Chacona it has been proposed that these structures were built between 1854 and 1881, showing a solstitial orientation that could be explained by Masonic symbolism, since the then property owner was a known

---

16 Juan de Abreu Galindo, *Historia de la conquista de las siete islas de Canaria*, ed. Imprenta Lithografia y Libreria Isleña, Miguel Miranda (Santa Cruz de Tenerife, 1848): p. 175.

17 Antonio Tejera Gaspar, José Juan Jiménez González, and Jonathan Allen, 'Las Manifestationes Artísticas Prehispánicas y su Huella', *Historia Cultural del Arte en Canarias* (2008), Vol. I: pp. 195–96.

18 J. A. Belmonte, A. Aparicio, C. Esteban, 'A Solstitial Marker in Tenerife: The 'majanos de Chacona', *Archaeoastronomy*, no. 18 (*JHA* Vol. 24), (1993), p. 65; Esteban, 'Arqueología soñada'.

19 Esteban, 'Arqueología soñada'.

freemason.[20] However, the large number of Maroiço structures in Pico Island does not seem to be compatible with such an explanation.

Case Study 2: The Artificial Caves from Monte Brasil in Terceira Island

Maritime traders returning from India and from the American continent used the Azores islands, and especially Terceira, as strategic harbours. Protection was assured from 1521 onwards by Portuguese fleets. With the increase of privateer activities, there was the need for permanent fleet defence units that would prevent entry into the mainland. In the middle of sixteenth century the construction of S. Brás fortress in S. Miguel Island started, followed by plans to build S. Sebastião fortress in Terceira Island. The construction of S. João Baptista fortress, also in Terceira Island, eventually started already under the rule of Filipe II from Spain, at the end of sixteenth century, in the isthmus of Monte Brasil peninsula.[21]

The Caves of Monte Brasil
In Monte Brasil, five rock-cut caves are commonly attributed to having had a military or civil purpose.[22] In this case study we will briefly analyse three of these caves situated in two coastal locations on the west face of Monte Brasil (Fig. 12 and Fig. 13).

---

20 Antonio Aparicio, and César Esteban, *Las pirámides de Güímar: mito y realidad* (Centro de la Cultura Popular Canaria, 2006); Aparicio and Esteban, 'Sobre la possible influencia del simbolismo masónico en las orientaciones de las Morras o "Pirámides" de Chacona, en Güímar'.

21 Nestor de Sousa, 'Programas de arquitectura militar quinhentista em Ponta Delgada e Angra do Heroísmo. Italianos, italianização e intervenções até ao século XVIII: a ermida de S. João Batista na fortaleza do Monte Brasil', *ARQUIPÉLAGO. História*, 2ª série, Vol. 6 (2002): pp. 53–224

22 Ribeiro et al., 'New Unknown Archaeological data in Azores'.

Figure 12 Location of Sites 1 and 2 in Monte Brasil, Terceira Island.

Figure 13 Sites 1 and 2 in Monte Brasil.

122 Land, Sea and Skyscape: Two Case Studies of Man-made Structures...

The three caves have locations that present a distant horizon between SW and NE, with the islands of Pico and S. Jorge being visible in the West far horizon (Fig. 14 and 15).

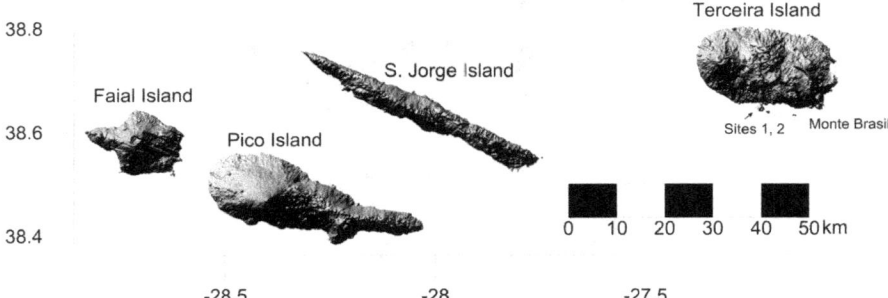

Figure 14 Islands of Pico and S. Jorge visible to the west of the sites (axis in degrees of latitude and longitude).

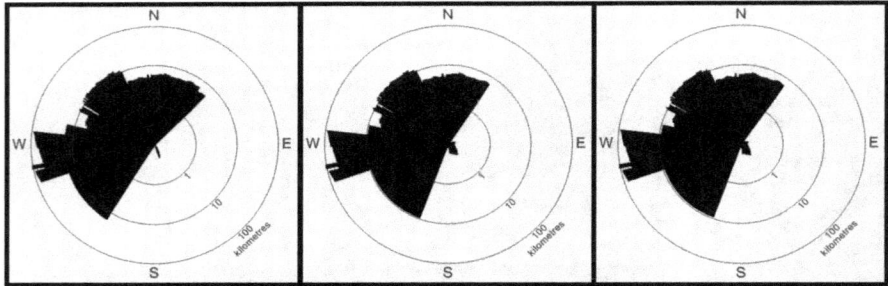

Figure 15 Visibility (in black) from the three caves: (left) Cave 1. (middle) Cave 2. (right) Cave 3.

The cave in site 1 measures 5m x 4m with a small water basin hewn into the back wall. There are two water collecting channels, one on each side of the basin (Fig. 16).

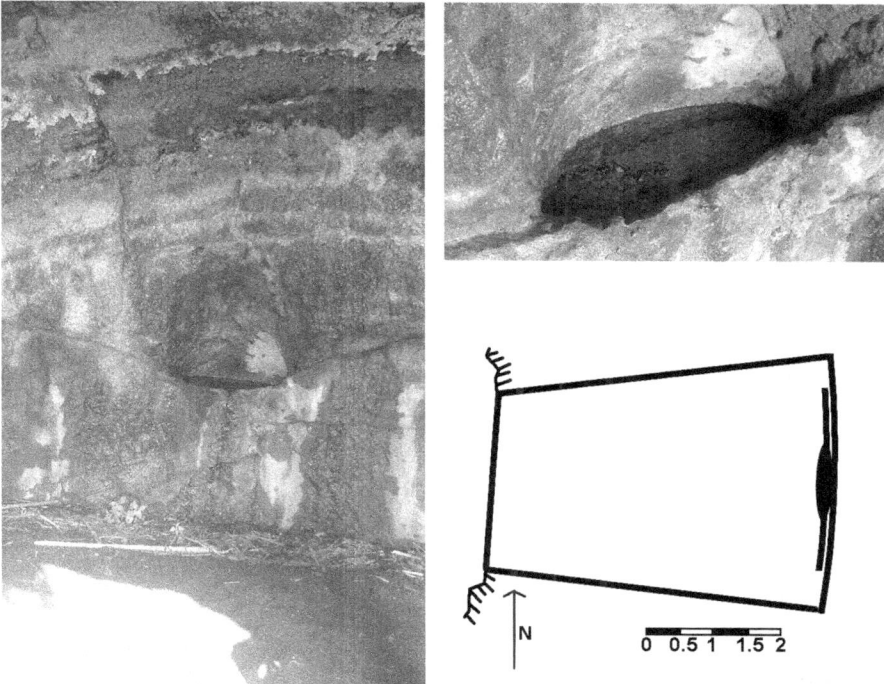

Figure 16 Site 1 cave: (left) cave with a small water basin; (top right) detail of the water basin; (bottom right) plan of the cave.

An interesting light effect can be observed at equinox sunset: the entrance projects a shaft of light towards the back wall, framing the basin at the centre. Figure 17 shows a photograph and a diagram of this effect, twenty minutes before sunset (with the sun at an azimuth of 267°). At sunset, on a cloudless sky, the light projection would be centred on the back wall. This effect would only be visible for about two days around the equinoxes, if one accepts that an offset of about one degree would be visually noticeable.

124 Land, Sea and Skyscape: Two Case Studies of Man-made Structures...

Figure 17 (top left) Photo taken on autumn equinox 2013, twenty minutes before sunset; (top right) light-shaft diagram for equinox 2013, twenty minutes before sunset; (bottom) Sun setting over S. Jorge Island on autumn equinox 2013 as seen from Cave 1 (Pico Island at left, S. Jorge Island in the middle and Pico at left).

In the second site, located one hundred metres to the south of the first, there are three caves, two of nearly equal dimensions and a third smaller cave located in a less accessible position (Fig. 18).

Figure 18 The three caves from site 2.

Cave 2, the one on the left in Fig. 18, has a 6.7° slope with a 270° aspect and a channel around the walls that connects to a second, external, channel. The combination of these elements would ensure that any liquid poured inside would flow out. Four basins in an asymmetrical arrangement are connected to the internal channel (Fig. 19).

Cave 3, the middle one in Fig. 18, has the same dimensions as cave 2, however its inner chamber is very different. It features several steps down to a permeable tank, with a possible seat feature at the bottom and along the walls, as well as a niche in the right corner of the back wall (Fig. 20).

Just as for site 1, some interesting illumination effects take place at sunset, especially at the equinoxes, when the light projection in the back wall touches the left-hand side wall of cave 3 (Fig. 21, left), and the left borders of the leftmost basins in cave 2 (Fig. 21, right). Three months later, at summer solstice sunset, the light-shaft would reach the rightmost basin in cave 2 (Fig. 22 b).

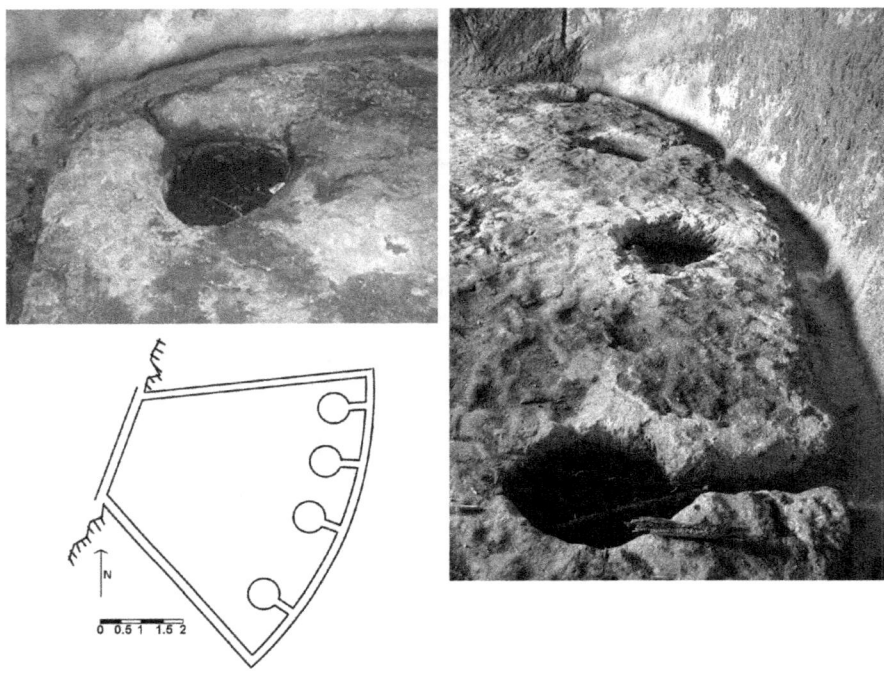

Figure 19 Cave 2 of site 2: (top left) detail of one basin; (bottom left) plan of the cave, with basins and channels; (right) the four basins at the back wall of the cave.

Figure 20 Cave 3 of site 2: (centre) steps down to the tank and a possible seat; (left) plan of the cave with steps, seat feature and niche; (right) the niche in the back wall.

Figure 21 Sunlight projection twenty minutes before equinoctial sunset in: (left) cave 3; and (right) cave 2.

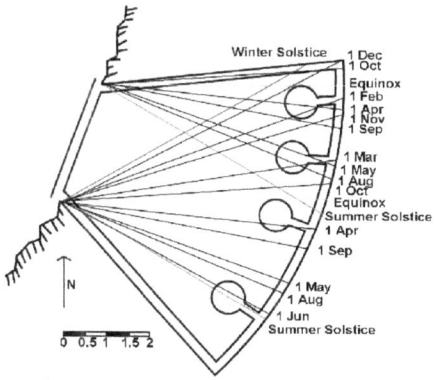

Figure 22 a) Diagram of sunset projected light-shafts in cave 2, for several dates along the year. The projection moves clockwise from Winter solstice to Summer solstice and then anti-clockwise back to Winter solstice. During spring and summer the projection is in the back wall.

Figure 22 b) (left) Sunset illumination in cave 2, thirteen minutes before sunset, eleven days after summer solstice in 2013. **c)** (right) Sunset illumination of the leftmost wall in cave 2, minutes before winter solstice sunset in 2013.

From all these caves sunset at the equinoxes is seen over S. Jorge Island (Fig. 17, bottom, and Fig. 23). Just as a curiosity the niche in the right corner of the back wall from Cave 3 faces the highest point in Terceira Island, Serra de Santa Bárbara, 1021m high, at 322° of azimuth and 4.7° of horizon height, which corresponds to 41.5° of declination (Fig. 23). Such a northern declination is outside the possible positions of Sun, Moon or planets. The closest first magnitude star that had this declination was Capella around 200 CE.

Figure 23 Panorama generated for cave 3, with the trajectories of the Sun (in yellow) for winter solstice (left), equinox over S. Jorge Island (middle), and summer solstice (right). The peak of Santa Bárbara is seen at the left. The two islands from the central group are seen in the centre: Pico at left and S. Jorge in the middle.

Discussion
The purpose of these caves, when and who built them are not known today. Some older people, including military, informed us that from the beginning to middle of the twentieth century it was common to have people bathing in the tank from cave 3, because the water was said to have good properties for skin treatment. However, there are no signs of thermal water as no temperature increase was measured. The water from this tank is not drinkable today: it has very high levels of silica, chloride and sodium (the last two are unambiguously related to sea spray deposits inside). It has also high levels of sulphate, probably derived from volcanic degassing.

Biogenic stalactites in cave 3 were measured and compared with similar stalactites with the same composition from a nearby water cistern with a known construction date from the seventeenth century. Assuming linear growth and the same nutrients and ambient conditions, one can compare the two stalactites to arrive at an age estimate. This suggests that the stalactites from cave 3, longer than those of the water cistern, would date to between 2000 to 400 years ago. An attempt made to date this material by radiocarbon, gave a result post 0 BP, probably due to aquifer contamination by nuclear bomb tests in the 1950s, making this methodology useless.

We can speculate that these caves may have integrated some kind of Sun-related symbolism in their construction, but without any other written sources or archaeological evidence two questions remain without answer: when were those caves constructed and for what purpose?

Conclusion

This preliminary study has revealed elements embedded in the construction of the structures that required a selection process for their location and some kind of building conventions in their construction, particularly with respect to their orientation.

In the case of the Maroiços from Pico Island there is clear intentionality in their orientation. It is not possible to determine if this orientation is topographic in intent, that is whether its intent was to align with the mountain top, or archaeoastronomical, that is whether the summer solstice sunset over Faial Island was the primary intent, or indeed a combination of the two, a dual alignment. However, the geographical clustering in such a small area suggests a bidirectional, and highly symbolic, alignment, even if the main purpose of their construction might have been ecological.

In the case of the caves situated near the sea at Monte Brasil, in Terceira Island, the lack of cultural context (ethnographical, historical or archaeological) allows only speculation about the intentionality of their orientation in the direction of the three islands of Pico, S. Jorge and Faial. However, from a phenomenological point of view, the observed illumination effects at equinox, when the sun sets over S. Jorge Island, cannot be simply ignored. Had the users of these caves noticed these effects they might have incorporated them in their construction, function or symbolism.

In conclusion, whatever was the period of construction of the Maroiços and the Monte Brasil caves, the effort in their building, their architecture and patent relation with the landscape, seascape and skyscape is enough to justify further research, excavation, as well as interest from both the local and national authorities and the public.

Acknowledgments

Landscape treatment, including distance to horizon profiles, horizon height profiles and horizon marks was made using the panorama generated data produced by custom software developed by Andrew Smith (available in http://www.agksmith.net/horizon), to whom we are grateful, and an SRTM based modified DTM grid.

Other References
Costa, Susana Goulart and Deolinda Maria Adão, 'Discovery, Settlement, and Demographics', *Azores: Nine Islands, One History=Açores: Nove Ilhas, Uma Historia*, (Berkeley, CA: UC, Berkeley, Institute of Governmental Studies Press, 2008): pp. 209–27.
Ribeiro, Nuno, Anabela Joaquinito, and Sérgio Pereira, 'Phoenicians in the Azores, Myth or Reality?', Congress SOMA 2011, Catania, Italy. Mediterranean Archaeology. Abstracts.

Appendix
An archaeological survey during the summer of 2013 in Pico Island, provided some additional data on the Maroiços cluster discussed in the main text.[23] The material evidence found is summarized in this appendix.

On top of structure M112 a possible remnant of an ancient circular floor was detected. Seven basalt artefacts were retrieved from this area (Fig. A1). The corridor and chamber of the twenty metre long structure M7 was excavated. Figure A2 shows the plan and longitudinal cuts of this Maroiço. Several materials were retrieved from this excavation: metallic points and hooks; lithic artefacts probably used as fish net weights and a fragment of a stone edge; a bone artefact; shells and fragments of terrestrial fauna (Fig. A3). Finally, two coal samples, collected in two stratigraphic units inside the chamber, were also dated (Fig. A4):

• MAR7S1CUEIII, at about 40cm depth, was dated to 300 +/- 30 BP: Cal 1489CE to 1604CE (69.6%) AND Cal 1610CE to 1654CE (25.8%);

• MAR7S1AUEIV, at about 80cm depth, was dated to 230 +/- 30 BP: Cal 1530CE to 1538CE (0.9%) AND Cal 1635CE to 1684CE (44.7%) AND Cal 1736CE to 1805CE (39.2%) AND Cal 1935CE to post 1950 (10.6%)

The material record, while not explaining what may have been the original purpose of the chamber and corridor, indicate that they were, to a certain extent, used for fishing or hunting associated activities.

---

23 Ribeiro, et al., 'Estudo Histórico Arqueológico sobre as Construções Piramidais existentes no Concelho da Madalena da ilha do Pico (Açores)'.

Figure A1 Basalt artefacts found in Maroiço M112: a scraper, a flaked point with rope marks, a fishnet weight, a dormant grindstone and a polisher

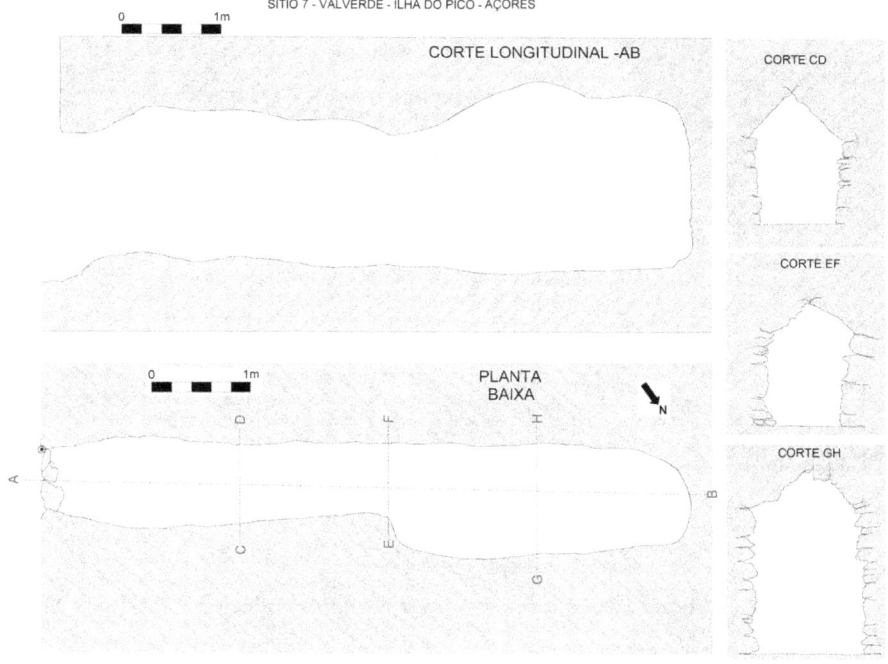

Figure A2 Plans and longitudinal cuts of the corridor and chamber of structure M7.

132 Land, Sea and Skyscape: Two Case Studies of Man-made Structures...

Figure A3 Materials collected from structure M7

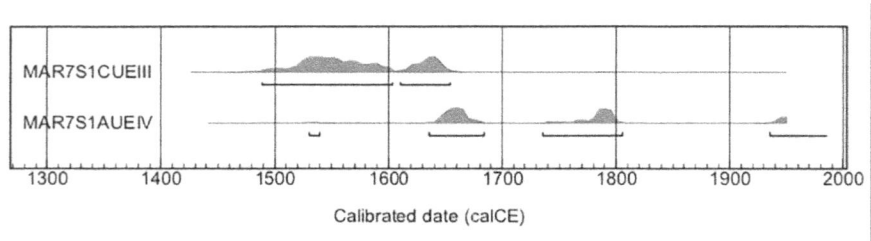

Figure A4 Calibrated dates for the two coal samples collected in Maroiço M7

Book Reviews

The Idea of Order in Review

Richard Bradley, *The Idea of Order: The Circular Archetype in Prehistoric Europe*. Oxford: Oxford University Press, 2012. £67 (cloth).

Any new book by Richard Bradley is eagerly awaited, not only because he is so respected as an archaeologist, but because his books are well researched, rich with pertinent examples and written in a style which manages to be both authoritative and accessible to the reader. His latest book, *The Idea of Order: The Circular Archetype in Prehistory*, is no exception. What is exceptional is that, in the wealth of literature that has been written about prehistory in general, and the Neolithic and Bronze Age in particular, he has managed to look at the period from a new angle, one which explores the prehistoric conception of space. There is a growing acceptance that space and place is important in the location of prehistoric buildings and reports of excavations increasingly locate finds within the landscape. Richard Bradley takes this an ambitious step further by examining patterns of dwellings and monuments throughout prehistoric Europe.

Bradley explores the idea of the circular archetype by drawing on many fine examples of circular monuments and dwellings constructed in prehistoric Europe. He links the use of this circular design to curvilinear motifs found on rock art and portable artefacts. He eschews Jung's specialised use of the term archetype and even the spiritual dimension suggested by Eliade and defines it as '"an original model or prototype" on which other examples are based' simply 'to suggest the ubiquity of circular forms in prehistoric architecture and the design of objects'. He further theorises that the choice of linear or curvilinear buildings gives rise to or is informed by distinctive perceptions of the world. Throughout the book he uses case studies to illustrate the phenomenon as a whole. The case studies are accompanied by clear and precise plans and drawings, which add clarity to the text.

To look at the complete range of dwellings and monuments in such a large area as prehistoric Europe with its regional differences of culture and terrain and to do this for a period stretching over millennia is a massive undertaking but Bradley's material is clearly organised into different sections and chapters so the reader is able to follow his purpose. He firstly introduces the subject of circular houses and looks at those parts of Europe where this 'circular perception of space' existed for the longest period. He then considers the architecture of domestic dwellings and relates them to the forms of larger monuments. By contrast he also looks at circular structures in areas where rectilinear architecture was the norm and finally looks at the period when the circular archetype was abandoned.

His approach shows how archaeological theory has developed. In the 1970s and 1980s he says that archaeologists focused on the distinction between mobility and sedentism and functionalists were concerned with subsistence, ecology, and the political economy. However the work of the anthropologist Mary Douglas had a major influence on archaeology theorists by her introduction of ethnography to explain the meanings and significance of domestic space. Through a gradual process culminating perhaps in Binford's (2001) *Constructing Frames of Reference*, which favours an 'analytical method for archaeology theory building, using ethnographic and environmental date sets', there has been a shift from a purely functional model to a broader spectrum which now considers the interpretation of circular structures in terms of systems of belief.

Bradley believes these theories can complement one another and he offers a mixture of practical, pragmatic reasons for circular dwellings alongside a cosmological approach to explain the worldview of these early builders. Stating that 'perceptions of space are as varied as other parts of human culture', he explains the differences between the perceptions of dwellers in open hilly country and those of woodland dwellers. In open country—and more particularly open country which has higher vantage points—most groups are able to view the entire dome of the sky and can identify where the earth meets the heavens and where both touch the sky. He continues by saying that as part of that process they become especially aware of the movements of the sun and moon. Without speculating on an exact belief he gives examples of monuments which are aligned to celestial movements, particularly sunrise or sunset at the solstices. Living within the environment of a circular horizon and under the circular dome of the sky the circular architecture reflects the builders' awareness of space. On the other hand, communities who live in woodland where the horizon and sky are obscured may be less aware of the heavenly bodies and their notions of space can be dominated by the importance of certain directions such as paths or trails. He concludes that two different ways of viewing space would have developed: for some it would be essentially linear, a world of directions and for others space would have been curvilinear extending to the horizon in a space where this circular configuration was mirrored by the dome of the sky. He goes on to cite examples where dwellings mirror their inhabitants' cosmology and says that the houses 'offer interpretations of the ideas on which the natural order depends'. He continues by quoting from Cauvin's (2000) *The Birth of the Gods and the Origins of Agriculture* to say that rectangles are rare in nature and represent that which has been fashioned and made concrete. Bradley's ideas on these two differing perceptions of space mark a breakthrough in ideas about prehistoric architecture and need to be borne in mind throughout the rest of the book.

Bradley then goes on to look in detail at the round dwellings and monuments which characterise much of prehistoric Europe, comparing them with rectilinear dwellings and long mounds, which are also widely distributed. He also looks at the pottery for the period and notices that the settlers in the rectilinear buildings associated with the Linear Pottery Culture of Neolithic Central and Northern Europe chose curvilinear designs with which to decorate their pottery and observes that the later metal artefacts were also decorated with circles and spirals. By contrast, in Britain and Ireland where the buildings were circular the Grooved Ware of the period is decorated with predominantly angular designs. Beaker pottery, adopted later, is also decorated by incised linear patterns, yet their main deposition was at circular sites. These regional differences are also found in the shapes of chambered cairns in Europe which are predominantly circular apart from Central Europe where they are rectangular. He says that the clear distinction and the contrast between rectilinear and curvilinear forms is rarely observed and more rarely discussed, yet it is found very widely. This book certainly attempts to fill this gap though the evidence is sometimes confusing. Bradley admits that sometimes there is a problem with chronology, as despite the similarities of design between dwellings and monuments they tended to have been built at different times as much as a century apart.

If there is a failing in the work, it is simply the scale of the area examined, the differences in construction and materials employed and the traditions that influenced them over the millennia spanned by prehistory. That houses tended to be built in timber, which has a limited life-span and leaves less of a footprint for archaeological excavation, whereas the tombs associated with them were built in stone, illustrates part of the problem.

The book looks at current archaeological method and theory, if only implicitly. Once hampered by the confines of the Three Age system, modern archaeological research looks more closely at transitions and what they mean both in practical, economic terms and also the effect on society and beliefs. The nature of archaeological excavation is particular and detailed and rarely applicable to sites other than the one being investigated, yet Bradley feels that 'some of the most useful observations have been neglected in the search for general principles'. Instead he uses his numerous and carefully described examples as building blocks towards a general picture from which he draws his conclusions. He suggests that the design of domestic dwellings was used as a prototype for monumental designs. That does not fit completely with the other evidence he presents of monuments often being directed towards the sun at the turning points of the year or that they occupied special places in the landscape. It would appear that while archaeologists increasingly take account of ethnography, archaeoastronomy, and sacred geography, there has not been a complete integration as yet.

His discussion of Uisneach in Ireland is pertinent at this point. Uisneach, which has been identified as the centre of Ireland, is a plateau-topped hill that dominates the view from all directions. It contains at least twenty ancient monuments including circular enclosures, a megalithic tomb, a round mound, a circular ditched enclosure, and a considerable number of small circular barrows or ring ditches. Despite saying that the circular archetype as described in the book does not equate to Eliade's definition, Bradley expresses the idea (without using Eliade's term *axis mundi*) that the monuments on Uisneach could have been microcosms of the land around it. Discussing the round tells in Romania, he refers to his earlier work, *The Significance of Monuments* (1998), by saying that 'in a sense they *ritualised* the features of settlements of a kind that had been occupied in the past and, in doing so, they invested these forms with a new significance'. Additionally while describing the roundels of Hungary, he suggests that the many possibilities for their form of construction could be the movements of the sun and the moon, the contrast between day and night and the passage of the seasons on which life depends.

Bradley continues his assessment by looking at the juxtaposition of circular and rectangular buildings saying that it seems unlikely that any one model can accommodate all the available evidence. By going from individual sites and looking at a larger pattern there is some evidence that they were sited at important parts of the landscape where routes met. Alternatively it could have been an expression of the demarcation of sacred and profane space. Additionally he thinks that the juxtaposition in the Netherlands and south Scandinavia shows a duality which implies that there was a rite of passage from the houses of the living to the houses of the dead. In Bronze Age Scandinavia there was a three-tier cosmology which explained the relationship of the land, the sea, and the sky. The sun rose from the water and travelled across the land until it set, then it was taken in a boat underneath the earth during the night to the place where it rose again in the morning. Circular motifs thought to represent the sun are also present there.

Bradley looks at how the use of circular buildings went out of use and associates this transition with outside influences. In Ireland these influences were Norse, and it is significant that the last to change were the Gaelic-speaking areas. He says it was slow to die out because the circular ordering of space as evidenced by the buildings retained its power. The images were respected and even renewed because they included elements that seemed important and that because the circular layout of the house was identified with a wider system of belief, it would have been difficult to adopt another style of dwelling.

Having examined all the evidence, he draws a map of prehistoric Europe showing the dual character expressed in the buildings. Circular buildings feature strongly in those regions which were connected by the sea. In Central and Northern Europe where there were more land connections, there was a preference for rectilinear buildings and round burial mounds. Thus there is a clear division between these two areas, and Bradley hints that this was because of the way the two areas adopted farming practices. Central and Northern Europe were colonised by farming communities whereas the coastal areas had more indigenous development. This would have led to a sense of regional identity which reinforced their forms, and it was only when their lands were taken over by Roman, Saxon, Viking, or Norman invaders that their circular worldview was overthrown.

In Richard Bradley's summing up, it is disappointing that he does not refer back to his innovative contrast between open landscape and wooded interiors, which in this reviewer's opinion underpins the detail of the book. Instead he leaves us with tantalising questions: Were the locations chosen because they conformed to a particular idea of order? Did the circular monuments epitomise a circular conception of space?

—Liz Henty
University of Wales, Trinity St David

Stonehenge decoded?

Mike Parker Pearson, *Stonehenge: Exploring The Greatest Stone Age Mystery*. London, New York: Simon & Schuster Press, 2012. £25 (cloth).

This book provides a popular account of the 'Stonehenge Riverside Project'—a seven year series of excavations between 2003 and 2009 in and around Stonehenge by a large team of archaeologists led by Mike Parker Pearson, and a summary of some of the discipline's recent relevant discoveries into our prehistory. The project was prompted by Parker Pearson's association with Ramilisonina, a southern Madagascan archaeologist of the Tandroy people who continue to build stone monuments. On being taken by Parker Pearson to Avebury stone circle, Ramilisonina saw no mystery as to their meaning, since a monument of stone must be a cenotaph to the ancestors, compared to monuments of wood which are for the living. Out of Parker Pearson's adoption of this interpretation has come the 'materiality model'—those monuments of wood and stone were the 'stations' for festivals of life and death. This model has been hailed by some of his colleagues as finally achieving the 'decoding' of Stonehenge.

It used to be thought that wooden monuments close to Stonehenge, Durrington Walls and Woodhenge, were earlier wooden dummy runs for the later upgrade to stone at sarsen Stonehenge. But once the Riverside team in 2008 dated Durrington Walls and the close-by Woodhenge to be contemporaneous with Stonehenge this 'lithification' thesis was dropped and favoured the materiality thesis. The model predicted that funereal feasting rituals conducted at Durrington Walls would have been a precursor to taking the processed remains of the illustrious dead to be interred at Stonehenge. The key to decoding Stonehenge was therefore its relationship to the wooden Durrington Walls. Parker Pearson predicted that if this model is correct then it predicted that Stonehenge was linked to the River Avon, that the wooden Durrington Walls had to have its own avenue linking it to the river upstream of Stonehenge, that the river route would be a funeral highway, and that evidence of burials should exist at Stonehenge. The many specialists in the research team included the archaeoastronomer Clive Ruggles, who suggested that Durrington Walls and Stonehenge had horizon alignments on the sun's solstices and the moon's standstills, and that these would have served calendrical ancestor ritual purposes, rather than those of a high precision observatory. This model therefore sees Stonehenge as one component of an integrated complex of monuments of a prehistoric sepulchral cosmology. Parker Pearson claims that exactly the same cosmology can be found at Avebury, since here just as at Stonehenge the Avebury stone circle is linked by the River Kennet to the wooden West Kennet Palisades and the Sanctuary. If correct it raises archaeological models beyond the limitations of site-exclusive excavation reports, and integrates its interpretations with anthropological analogy and archaeo-astronomical findings. And by a return to a hypothetico-deductive methodology without excluding post-processual insights, Parker Pearson set up a number of tests to his model.

In a series of excavations in and around Durrington Walls and the Avon riverside, the project team have discovered that that a large temporary village of huts and other buildings were sited early on to house thousands of people for monument building work and pig feasts, that an avenue of rammed flint does in fact connect that monument with the river, that along much of the River Avon's high banks a number of wooden towers were located alongside large cremation fires, that the Stonehenge Avenue did in fact connect with the River Avon and was marked by an early 'Bluestonehenge' later replaced with a henge when the bluestones were moved to add to Stonehenge, and that at Stonehenge for much of its over one thousand years of use it was a cremation cemetery. These are a remarkable series of discoveries that seem to confirm the materiality model, and the book is an accessible source for the latest archaeological knowledge of Stonehenge and our local prehistory. The book is written in the now familiar style of a publicity-savvy British archaeology, introducing by personal disclosure the characters who variously made up the changing team of the Riverside Project. And with the archaeological engagement with anthropology and archaeoastronomy this seems to finally signal an end to the destructive and acrimonious debates of the sixties and seventies that can allow a decoding of Stonehenge. However there are many reasons to be cautious while still welcoming this publication.

At the very beginning of his book Parker Pearson claims the scientific high ground—that a theory must withstand the tests of evidence and discoveries of new evidence. If a hypothesis cannot explain evidence then that theory must be rejected. There is another model of the scientific process, in which disciplinary and paradigm boundaries immunise scholars against evidence that is counter to their preferred interpretations. Rather than a scholarly landscape littered with defunct rejected theories this second model predicts that mutually incompatible theories survive by research groups ignoring the evidence and publications of others to the detriment of a broad front forward movement in knowledge. Parker Pearson claims to follow the first 'Popperian' model of scholarship yet displays classic symptoms of the second 'Kuhnian' practice. In spite of an impressive energy to get out and do the fieldwork, Parker Pearson has not considered all extant theories or evidence and that work has still to be done.

This is particularly clear with his adoption of archaeoastronomy. While it is refreshing to read a senior archaeologist at last engaging with archaeoastronomy, his justification for his adoption of it is worrying. He argues that archaeology has been under 'seige' from the earlier archaeoastronomers like Hawkins, Thom, and North, but can now be adopted because Ruggles is an archaeologist and therefore any astronomical properties of a monument can be grounded in 'archaeological knowledge'. This is an unjustified comment, since for example the work of John North is one of the most archaeologically detailed and painstaking examples of archaeoastronomy research ever made. Scholarship does not move forward through disciplinary amity relationships, but requires the sober and non-partisan assessment of evidence. With regard to Stonehenge and its 'astronomy' it would have paid Parker Pearson to pay more regard to John North's work. Parker Pearson claims, apparently on Ruggles' authority, that Stonehenge had axial horizon astronomy alignments on summer solstice sunrise and winter solstice sunset, the four station stones have alignments on the southern major moonrise, the northern major moonset and the summer solstice sunrise, and the entrance post holes on the northern major moonrise. None of these claims were first made by Ruggles and some are so problematical that they display a slipshod attitude to the architecture of the monument. The summer solstice sunrise claim was first made by Stukeley in the early eighteenth century, the station stones' alignments by Hawkins in 1963, and the entrance post holes northern major moonrise alignment by Newham in 1972. Why is Parker Pearson accepting the claims of Hawkins and Newham while simultaneously viewing them as enemies of archaeology? And why does he amalgamate them with John North? North showed in painstaking detail that these three claims in particular are not true to the archaeology of the monument or the field method requirements of archaeoastronomy. The close argumentation for this view has been in the public domain from North and others for at least two decades now, and one wonders why Parker Pearson is practising Khunian paradigm defence by bibliographic exclusion. The convincing picture made by North is a closely specified double alignment from the Heel Stone on winter solstice sunset and the southern minor standstill moonset, both through the Grand Trilithon, which together generate a dark moon at winter solstice every nineteen or so years. Since his own materiality model requires Stonehenge to be a cremation cemetery, then a ritual that coincides with the start of the longest darkest night makes far more sense than the start of the longest brightest day through his claimed summer solstice sunrise alignment.

While the archaeoastronomy is poorly advised, the archaeology is under-interpreted. While both Durrington Walls and Woodhenge are within his 'domain of the living', a male burial is found along the Durrington Walls avenue and a three year old girl child with her skull split in two was at the centre of Woodhenge. Similarly at the Sanctuary, the Avebury equivalent of Woodhenge, an adolescent child was buried and Aubrey Burl's review of the antiquarian evidence suggests that many more human remains were once there. Some dead are therefore found in the proposed 'the land of the living'. And at Stonehenge Mike Pitts has suggested that an oak lintel spanned the gap beneath the half-height stone 11 in the outer circle of sarsens. Together with the woodworking mortise and tenon joints that Parker Pearson acknowledges were used by the builders at Stonehenge, this suggests that the monument is not reducible to stone alone, just as 'the land of the living' has to include the dead. The reverse holds true, since Parker Pearson's excavation has revealed that the Durrington Walls avenue was lined with sarsen pillars. Similarly rather than the multi-phase roofed buildings favoured by most archaeologists in the past, we now know that both the Sanctuary and Woodhenge were open monuments of wood *and* stone. And while only 6% of the Avebury Circle has been excavated, nobody has suggested that its main function was as a cremation cemetery. Yet in spite of a pointedly different architecture, John North has shown it shares with Stonehenge the same axial alignments on the setting winter solstice suns and the southern standstill moonsets. The materiality model is insensitive to both the detailed astronomy and archaeology, and this anomalous evidence challenges scholars, including Parker Pearson, to be true to science and either amend or reject the theory.

The materiality model is the latest version of the sepulchral model which has been around since Petrie first suggested it in the 1880s. Archaeologists of the Neolithic have generally viewed monuments housing the elite dead as signifying the decisive break with forager primitiveness and the beginning of civilisation. Instead of seeing monument building and ancestor worship as foundational of culture, anthropology suggests that it is a revision and reversal to the forager cosmology that preceded it. Parker Pearson's use of one megalith building culture separated by thousands of miles and years as an analogy for Stonehenge is unsafe. It could be countered by others that would reverse the meaning of materiality, since wood that rots can be seen as the domain of the dead and resistant stone as the permanence of society. Since we now know that the Neolithic monument building cultures of NW Europe were cattle herders who continued to hunt and forage, then a more appropriate analogy would be to the global ethnographies of pastoral cultures. All suggest that young men only gain marital rights through gifts of cattle to wife givers, and that they receive these cattle from senior agnates. Ancestor worship is therefore a public demonstration of respect to them, since the elite dead are those senior agnates that are the source of cattle. But as the descendants of native hunter and gatherers, these nouveaux-riche cattle owners had to contend with the tensions of traditional forager bride-service obligations which were gradually eclipsed by bride-price negotiations. Seen as a simultaneous continuation and reversal of an earlier cosmology, the regionally competitive dynamic of monumental architecture can arguably be interpreted as addressing the builders' immediate concerns to overcome fraying clan loyalties and to signal externally their ability to mobilise labour in large handicap displays. This alternative anthropological model, not considered by Parker Pearson, predicts that our ancestors used a complex swan-song of allusions, their horizon alignments displacing earlier lunar phased rituals with an elite male solarising religion.

This is a book that must be read by all interested in the period and the issues. As one attempt to use a multi-disciplinary method to decode Stonehenge, it has failed, but linking archaeology with archaeoastronomy and anthropology clearly has much to offer.

—Lionel Sims
University of East London

NOTES ON CONTRIBUTORS

Pamela Armstrong carried out her research at the Sophia Centre for the Study of Cosmology in Culture, University of Wales Trinity Saint David. She achieved a distinction for MA dissertation, which considered the skyscapes of ancient Britain —in particular, the archaeoastronomic properties of monumental pre-historic architecture. Her research focused on the Mesolithic to Neolithic transition in western England. Her study is ongoing as she broadens the scope of the material record explored to include not just stone chambered tombs, but a wide range of monuments and structures from that time, including passage tombs, earth works and stone circles. She has presented papers at Theoretical Archaeology Group conferences and the National Astronomy Meeting, and is a contributor to the *Journal of Skyscape Archaeology* and the *Journal of Physics: Conference Series*.
yewcedditcyster@btinternet.com

Daniel Brown is a physicist who developed hands-on astronomy teaching, while at the Ruhr-Universitaet Bochum, Germany. He became interested in archaeo-astronomy working with Prof W Schlosser, and is a founding member of the Initiativekreis Horizontastronomy im Ruhrgebiet. He has a doctorate in theoretical stellar evolution and currently lectures in astronomy at Nottingham Trent University. His work combines the outdoor classroom, sustainability, place experience and archaeoastronomy as exemplified by his forthcoming publication 'Skyscapes Present and Past—From Sustainability to Interpreting Ancient Remains' in F Silva and N Campion (eds) *Skyscapes: The Role and Importance of the Sky in Archaeology* (Oxbow, 2015). His practical work in the Peak District National Park focuses upon Gardom's Edge as well as Light pollution education.
daniel.brown02@ntu.ac.uk

Antonieta Costa has a PhD in Social Psychology of Organizations (ISCTE, Lisbon Univeristy Institute). She is currently a post-doctoral researcher at the University of Porto, and a researcher of CITCEM (Trans-disciplinary Research Centre: Culture, Space and Memory) where she focuses on the roots of popular culture, archetypes and myths.
antonieta_c@hotmail.com

Liz Henty left her accountancy career to take the Cultural Astronomy and Astrology MA at the University of Wales Trinity Saint David, where she achieved a distinction for her dissertation entitled 'An Examination of Possible Solar, Lunar and Stellar Alignments at the Recumbent Stone Circles of North-East Scotland'. After taking some short archaeology courses at Aberdeen University, she is now a PhD Student at the University of Wales Trinity Saint David, researching the divide between the disciplines of archaeology and archaeoastronomy. She has presented papers at SEAC and the Theoretical Archaeology Group conferences, is a

contributor to the volume *Skyscapes: The Role and Importance of the Sky in Archaeology*, edited by F Silva and N Campion (Oxbow, 2015) and co-editor of the *Journal of Skyscape Archaeology* (Equinox Publishing).
lizhenty@f2s.com

Anabela Joaquinito is an archaeologist with the Portuguese Association of Archaeological Investigation (APIA) and PhD candidate in Prehistory and Ancient History at the University of Salamanca. She has authored articles on the pre-Portuguese occupation in Azores islands, Mesolithic lithic artifacts and rock art.
anabela.joaquinito@apia.pt

Tore Lomsdalen was an executive and general manager within the international hotel industry with a certificate exam from the Faculty for Astrological Studies in London. Due to his special interest in the history of astrology, he enroled in the MA in Cultural Astronomy and Astrology at the Sophia Centre, University of Wales Trinity Saint David. Throughout this period he discovered an passion for archaeoastronomy and ultimately wrote his dissertation on the prehistoric Maltese Temples, for which he got a distinction. He is the author of *Sky and Purpose in Prehistoric Malta: Sun, Moon, and Stars at the Temples of Mnajdra* (Sophia Centre Press, 2014). He is currently continuing his research of Maltese prehistory, particualrly with respect to cosmological transitions.
tore.lomsdalen@gmail.com

Fernando Pimenta lectured on statistics at the Instituto Superior Técnico (Lisbon) and has thirty years of experience developing control systems for the industry and software systems for power quality and fault location in electrical energy high voltage grids. His interests are in cultural astronomy, and he has focused on the orientation of archaeological structures in the landscape. He is a member of Portuguese Association of Archaeological Investigation (APIA) and was co-president of the Local Organizing Committee of SEAC nineteenth annual meeting in Évora, Portugal. Two of his articles, one on 'Astronmy and Navigation' and another, co-authored, on the 'Megalithic Cromlechs of Iberia', were recently published on the *Handbook of Archaeoastronomy and Ethnoastronomy* (ed. Clive Ruggles, Springer, 2015).
fernando.pimenta@apia.pt

Olwyn Pritchard worked as an archaeologist before becoming a bum and general ne'er do well in the west of Ireland, where she found a battered copy of Alexander Thom's *Megalithic Sites in Britain* in a second hand shop. This inspired her to start researching the astronomical aspects of the many local stone rows, circles and dolmens. After returning to Wales, and raising her children, she studied archaeology at the University of Wales Trinity Saint David. Her dissertation featured the astronomy of an apparently anomalous north-facing dolmen in south Pembrokeshire. Olwyn then undertook a MA in Landscape Archaeology at the

same establishment, researching the land and skyscapes around Strata Florida, from 2000 BCE to 1200 CE. She has presented archaeoastronomy papers at conferences and is a contributor to the volume *Skyscapes: The Role and Importance of the Sky in Archaeology*, edited by F Silva and N Campion (Oxbow, 2015).
olwynepritchard@gmail.com

Nuno Ribeiro has a PhD in Prehistory and Ancint History from the University of Salamanca. He has worked as an archaeologist since 1995, coordinating more than one hundred archaeological works. He is President of the Portuguese Association of Archaeological Research (APIA) and author of several hundred publications in newpapers and academic journals in Portugal, Italy, Spain, Germany, United States of America and Brazil.
nuno.ribeiro@apia.pt

António Félix Rodrigues is Professor of Environmental Engineering and Environmental Sciences at the University of the Azores. His research interests include physics, mathematics, environmental sciences, communication sciences, landscape analysis and history. He has presented, by himself or in collaboration with co-authors, more than one hundred scientific communications at national and international meetings. He has published books on environmental sciences, communication sciences, and history.
rodriguessame@gmail.com

Fabio Silva is a NERC Research Associate at the Institute of Archaeology, University College London (UK) and a tutor in the Sophia Centre for the Study of Cosmology in Culture (University of Wales Trinity Saint David, UK), where he is responsible for a postgraduate taught module titled 'Skyscapes, Cosmology and Archaeology'. His current research interests focus on how humans perceive their environment (skyscape and landscape) and use that knowledge to time and adjust their social and productive behaviours.
fabio.silva@ucl.ac.uk

Lionel Sims is Emeritus Head of Anthropology at the University of East London. Published many papers combining anthropology, archaeology, archaeoastronomy and Indo-European poetics on interpreting prehistoric monuments with particular reference to Stonehenge and Avebury; a film of his research was made by National Geographic; member of the Stonehenge Round Table hosted by English Heritage and the Avebury Sacred Sites Forum hosted by the National Trust; Vice President of the European Society for Astronomy in Culture.
lionel.sims@btinternet.com

BACK ISSUES OF CULTURE AND COSMOS
Available from *Culture and Cosmos*, PO Box 1071, Bristol BS99 1HE, U.K.
E mail culture@caol.demon.co.uk for availability and prices.

Contents, Vol. 1 no 1 (spring/summer 1997)
Robin Heath: *An Astronomical Basis for Solar Hero Myths;* **Norris Hetherington**: *Ancient Greek Cosmology and Culture: a Historiographical Review;* **Alan Weber**: *The Development of Celestial Journey Literature, 1400 - 1650;* **Ken Negus**: *Kepler's Tertius Interveniens;* **John Durant** and **Martin Bauer**: *British Public Perceptions of Astrology: an Approach from the Sociology of Knowledge.*

Contents Vol. 1 no 2 (autumn/winter 1997)
Otto Neugebauer: *On the History of Wretched Subjects;* **Nick Kollerstrom**: *The Star Zodiac of Antiquity;* **Robert Zoller**: *The Hermetica as Ancient Science;* **Edgar Laird**: *Christine de Pizan and Controversy Concerning Star Study in the Court of Charles V;* **Jürgen G.H. Hoppman**: *The Lichtenberger Prophecy and Melanchthon's Horoscope for Luther;* **Elizabeth Heine**: *W.B.Yeats: Poet and Astrologer.*

Contents Vol. 2 no 1 (spring/summer 1998)
J. McKim Malville and **R. N. Swaminathan:** *People, Planets and the Sun: Surya Puja in Tamil Nadu, South India;* **Carlos Trenary:** *Yaxchilan Lintel 25 as a Cometary Record;* **Graziella Federici Vescovini:** *Biagio Pelacani's Astrological History for the Year 1405;* **Frank McGillion:** *The Influence of Wilhelm Fliess' Cosmobiology on Sigmund Freud;* **Nicholas Campion:** *Sigmund Freud's Investigation of Astrology.*

Contents Vol. 2 no 2 (autumn/winter 1998)
Giuseppe Bezza: *Astrological Considerations on the Length of Life in Hellenistic, Persian and Arabic Astrology;* **Angela Voss:** *The Music of the Spheres: Marsilio Ficino and Renaissance harmonia*; **Robert Zoller:** *Marc Edmund Jones and New Age Astrology in America.*

Contents Vol. 3 no 1 (spring/summer 1999)
Michael R. Molnar: *Firmicus Maternus and the Star of Bethlehem*; **Roger Beck:** *The Astronomical Design of Karakush, a Royal Burial Site in Ancient Commagene: an Hypothesis*; **Chantal Allison:** *The Ifriqiya Uprising Horoscope from* On Reception *by Masha'alla, Court Astrologer in the Early 'Abassid Caliphate.*

Contents Vol. 3 no 2 (autumn/winter 1999)
Robin Waterfield: *The Evidence of Astrology in Classical Greece;* **Remo Catani:** *The Polemics on Astrology 1489-1524*; **Claudia Rousseau**: *An Astrological Prognostication to Duke Cosimo de Medici of Florence.*

Contents Vol. 4 no 1 (spring/summer 2000)
Patrick Curry: *Historical Approaches to Astrology*; **Edgar Laird:** *Heaven and the Sphaera Mundi in the Middle Ages*; **George D. Chryssides:** *Is God a Space Alien? The Cosmology of the Raëlian Church.*

Contents Vol. 4 no 2 (autumn/winter 2000)
David J. Ross: *The Bird, The Cross, and the Emperor: Investigations into the Antiquity of The Cross in Cygnus*; **Angela Voss:** *The Astrology of Marsilio Ficino: Divination or Science?*; **Patrick Curry:** *Astrology on Trial, and its Historians: Reflections on the Historiography of 'Superstition'.*

Contents Vol. 5 no 1 (spring/summer 2001)
Demetra George: *Manuel I Komnenos and Michael Glykas: A Twelfth-Century Defence and Refutation of Astrology,* Part I; **Richard L. Poss:** *Stars and Spirituality in the Cosmology of Dante's Commedia.*

Contents Vol. 5 no 2 (autumn/winter 2001)
Arkadiusz Sołtysiak: *The Bull of Heaven in Mesopotamian Sources*; **Demetra George:** *Manuel I Komnenos and Michael Glykas: A Twelfth-Century Defence and Refutation of Astrology,* Part 2; **Garry Phillipson** and **Peter Case:** *The Hidden Lineage of Modern Management Science: Astrology, Alchemy and the Myers-Briggs Type Indicator.*

Contents Volume 6 Number 1 (spring/summer 2002)
Ari Belenkyi: *A Unique Feature of the Jewish Calendar - Dehiyot*; **Demetra George:** *Manuel I Komnenos and Michael Glykas: A Twelfth-Century Defence and Refutation of Astrology,* Part 3; **Germana Ernst** : *The Sky in a Room: Campanella's Apologeticus in defence of the pamphlet* De siderali fato vitando; **Tommaso Campanella:** *Apologia for the opuscule on* De siderali fato vitando.

Contents Volume 6 Number 2 (autumn/winter 2002)
Jesse Krai: *Rheticus' Poem* 'Concerning the Beer of Breslau and the Twelve Signs of the Zodiac'; **Anna Marie Roos:** *Israel Hiebner's Astrological Amulets and the English Sigil War*; **Nicholas Campion:** *Surrealist Cosmology: André Breton and Astrology.*

Contents Volume 7 Number 1 (spring/summer 2003) GALILEO'S ASTROLOGY
Nick Kollerstrom: *Foreword: Galileo as Believer*; **Nicholas Campion**: *Introduction: Galileo's Life and Work*; **Antonio Favaro**: *Galileo, Astrologer*; **Germana Ernst**: *Astrology and Prophecy in Campanella and Galileo*; **Nick Kollerstrom**; *Galileo as an Astrologer: Antonino Poppi: On Trial for Astral Fatalism: Galileo Faces the Inquisition*; **Guiseppe Righini**:*Galileo's Horoscope for Cosimo II de Medici*; **Mario Biagioli**: *An Astrologico-Dynastic Encounter; Galileo's Correspondence*; *Galileo's Letter to Dini, May 1611*; *On the Character of Sagredo: Galileo's judgements upon his nativity*; *Galileo's Horoscopes for his Daughters*; *Rome, 1630*; **Bernadette Brady**: *Four Galilean Horoscopes: An Analysis of Galileo's Astrological Techniques*; *A Sonnet by Galileo.*

Contents Volume 7 Number 2 (autumn/winter 2003)
Günther Oestmann: *Tycho Brahe's Geniture*; **Bernard Eccles**: *Astrological physiognomy from Ptolemy to the present day*; **James Brockbank**: *Planetary signification from the second century until the present day*; **Julia Cleave**: *Ficino's Approach to Astrology as Reflected in Book VII of his Letters.*

Contents Volume 8 No 1/2 (spring/summer autumn/winter 2004)
Valerie Shrimplin *Organising INSAP*; **Rolf Sinclair** *Foreword: INSAP IV in Oxford: A Summary*; **Nicholas Campion** *Introduction: The Inspiration of Astronomical Phenomena*:

Hubert A. Allen, Jr. *Hawkins' Way: Remembering Astronomer Gerald S. Hawkins*; Hubert A. Allen, Jr. and Terry Edward Ballone *Star Imagery in Petroglyph National Monument*; Mark Butterworth *Astronomy and the Magic Lantern*; Ann Laurence Caudano *Sun, Moon, and Stars on Kievan Rus Jewellery ($10^{th} - 13^{th}$ Centuries)*; Nicholas Campion *The Sun is God;* Anne Chapman-Rietschi *Cosmic Gardens*; Deborah Garwood *Paris Solstice*; N. J. Girardot *Celestial Worlds In the Work of Self-Taught Visionary Artists With Special Reference to Howard Finster's Vision of 1982*; John G. Hatch *Desire, Heavenly Bodies, and a Surrealist's Fascination with the Celestial Theatre*; Holly Henry *Bertrand Russell in Blue Spectacles: His Fascination with Astronomy*; Ronald Hicks *Astronomy and the Sacred Landscape in Irish Myth*; Chris Impey *Why Are We So Lonely?*; Bernd Klähn *The Aberration of Starlight and/in Postmodernist Fiction*; Nick Kollerstrom *How Galileo dedicated the moons of Jupiter to Cosimo II de Medici*; Arnold Lebeuf *Dating the five Suns of Aztec cosmology*; Andrea D. Lobel *Trailing the Paper Moon: Astronomical Interpretations of Exodus 12:1-2*; Stephen C. McCluskey *Wordsworth's 'Rydal Chapel' and the Astronomical Orientation of Churches*; David Madacsi *Sky: Atmospheres and Aesthetic Distance in Planetary and Lunar Environments*; Daniel R. Matlaga *A Journey of Celestial Lights: The Sky as Allegory in Melville's Moby Dick*; Paul Murdin *Representing the Moon*; R. P. Olowin *Robinson Jeffers: Poetic Responses to a Cosmological Revolution*; David W. Pankenier *A Brief History of Beiji (Northern Culmen)*; Richard Poss *Poetic Responses to the Size of the Universe: Astronomical Imagery and Cosmological Constraints*; Barbara Rappenglück *The material of the solid sky and its traces in cultures*; Brad Ricca *The Night of Falling Stars: Reading the 1833 Leonid Meteor Storm*; Patricia Ricci *Lux ex Tenebris: Etienne-Louis Boullée's Cenotaph for Sir Isaac Newton*; Sarah Richards *Die Planetentheorie: its uses and meanings for the Saxon mining communities and the culture of the Dresden Court 1553-1719*; William Saslaw and Paul Murdin *The Double Apollos of Istrus*; Petra G. Schmidl *Dusk and Dawn in Medieval Islam; On the Importance of Twilight Phenomena with Some Examples of Their Representations in Texts and on Instruments*; Valerie Shrimplin *Borromini and the New Astronomy: the elliptical dome*; Joshua Stein *Cicero's Use of Astronomy as Proof of the Existence of the Gods*; Antje Steinhoefel *Art and Astronomy in the Service of Religion:Observations on the Work of John Russell (1745-1806)*; Burkard Steinrücken *An interpretation of the `Sky Disc of Nebra' as an icon for a bronze age planetarium mechanism with parallels to the moving world-soul in Plato's* Timaeus; Gary Wells *Daumier and The Popular Image of Astronomy.*

Contents Vol. 9 no 1 (Spring/Summer 2005)
Gennadij Kurtik and Alexander Militarev *Once more on the origin of Semitic and Greek star names:an astronomic-etymological approach updated*; Prudence Jones *A Goddess Arrives: Nineteenth Century Sources of the New Age Triple Moon Goddess*; Louise Curth *Astrological Medicine and the Popular Press in Early Modern England.*

Contents Vol. 9 no 2 (Autumn/Winter 2005)
Marinus Anthony van der Sluijs *A Possible Babylonian Precursor to the Theory of ecpyrōsis*; Liz Greene *Did Orphic Beliefs Influence the Development of Hellenistic Astrology?*; Ariel Cohen *Astronomical Luni-Solar Cycles and the Chronology of the Masoretic Bible*; Tayra Lanuza-Navarro *An Astrological Disc from the Sixteenth Century*; J.C. Holbrook *Celestial Navigators and Navigation Stories.*

Contents Vol. 10 no 1 and 2 (Spring/Summer, Autumn/Winter 2006)

Culture and Cosmos

Lucia Dolce *Introduction: The worship of celestial bodies in Japan: politics, rituals and icons*; Lucia Dolce *The State of the Field: A basic bibliography on astrological cultic practices in Japan*; Hayashi Makoto *The Tokugawa Shoguns and Yin-yang knowledge (onmyōdō)*; John Breen *Inside Tokugawa religion: stars, planets and the calendar-as-*method; Mark Teeuwen *The imperial shrines of Ise:An ancient star cult?*; Lilla Russell-Smith *Stars and Planets in Chinese and Central Asian Buddhist Art from the Ninth to the Fifteenth Centuries*; Matsumoto Ikuyo *Two Mediaeval Manuscripts on the Worship of the Stars from the Fujii Eikan Collection*; Tsuda Tetsuei *The Images of Stars and Their Significance in Japanese Esoteric Buddhist Art*; Meri Arichi *Seven Stars of Heaven and Seven Shrines on Earth: The Big Dipper and the Hie Shrine in the Medieval* Period; Gaynor Sekimori *Star Rituals and Nikko Shugendô*; Meri Arichi *The front cover image: Myōken Bosatsu.*

Contents Vol. 11 no 1 and 2 (Spring/Summer, Autumn/Winter 2007)
Micah Ross *A Survey of Demotic Astrological* Texts; Francis Schmidt *Horoscope, Predestination and Merit in Ancient Judaism*; Stephan Heilen *Ancient Scholars on the Horoscope of Rome*; Joanna Komorowska *Philosophy among Astrologers* ; Wolfgang Hübner *The Tropical Points of the Zodiacal Year and the* Paranatellonta *in Manilius' Astronomica*; Aurelio Pérez Jiménez *Hephaestio and the Consecration of Statues*; Robert Hand *Signs as Houses (Places) in Ancient Astrology*; Dorian Gieseler Greenbaum *Calculating the Lots of Fortune and Daemon in Hellenistic Astrology*; Susanne Denningmann *The Ambiguous Terms ἑῴα and ἑσπερία, ἀνατολή, and ἑῴα and ἑσπερία δύσις* Joseph Crane *Ptolemy's Digression: Astrology's Aspects andMusical Intervals*; Giuseppe Bezza *The Development of an Astrological Term – from Greek* hairesis *to Arabic* ḥayyiz; Deborah Houlding *The Transmission of Ptolemy's Terms: An Historical Overview, Comparison and Interpretation.*

Contents Vol. 12 no 1 (Spring/Summer 2008)
Liz Greene *Is Astrology a Divinatory System?*; James Maffie *Watching the Heavens with a 'Rooted Heart': The Mystical Basis of Aztec Astronomy*; J.C. Holbrook *Astronomy and World Heritage.*

Contents Vol. 12 no 2 (Autumn/Winter 2008)
Mark Williams *Astrological Poetry in late medieval Wales: the case of Dafydd Nanmor's 'To God and the planet Saturn'*; Scott Hendrix *Choosing to be Human: Albert the Great on Self Awareness and Celestial Influence*; Graham Douglas *Luis Vilhena and the World of Astrology.*

Contents Vol. 13 no 1 (Spring/Summer 2009)
Josefina Rodríguez-Arribas *Astronomical and Astrological Terms in Ibn Ezra's Biblical Commentaries: A New Approach*; Andrew Vladimirou *Michael Psellos and Byzantine Astrology in the Eleventh Century*; Marinus Anthony van der Sluijs *The Dragon of the Eclipses—A Note*; Patrick Curry *Response to Liz Greene's 'Is Astrology a Divinatory System?'*

Contents Vol. 13 no 2 (Autumn/Winter 2009)
Liz Greene *Mystical Experiences Among Astrologers*; Peter Pesic *How the Sun Stood Still: Old English Interpretations of Joshua and the Leap Year*; Doina Ionescu *Virginia Woolf and Astronomy*; Carlos Ziller Camenietzki and Luis Miguel Carolino *Astrologers at*

War: Manuel Galhano Lourosa and the Political Restoration of Portugal, 1640–1668; **Nick Campion** *Astrology's Role in New Age Culture: A Research Note*

Contents Vol. 14 no 1 and 2 (Spring/Summer, Autumn/Winter 2010)
Dorian Gieseler Greenbaum *Introduction*; **Friederike Boockmann** *Johann Kepler's Horoscope Collection*; **J. Cornelia Linde (trans.)** *Helisaeus Röslin's Delineation of Kepler's Birthchart, 1592*; **J. Cornelia Linde and Dorian Greenbaum (trans.)** *David Fabricius and Kepler on Kepler's Personal Astrology, 1602*; **Dorian Greenbaum (trans.)** *Kepler's Delineation of his Family's Astrology*; **J. Cornelia Linde and Dorian Greenbaum (trans.)** *Kepler and Michael Mästlin on their Son's Nativities, 1598*; **J. Cornelia Linde and Dorian Greenbaum (trans.)** *Kepler's Methods of Astrological Interpretation for Rudolf II, 1602*; **J. Cornelia Linde and Dorian Greenbaum (trans.)** *Kepler's Astrological Interpretation of Rudolf II by Traditional Methods, 1602*; **J. Cornelia Linde and Dorian Greenbaum (trans.)** *Kepler's Letter to an Official on Rudolf II and Astrology, 1611*; **J. Cornelia Linde and Dorian Greenbaum (trans.)** *Excerpts from Kepler's Correspondence and Interpretation of Wallenstein's Nativity, 1624-1625*; **J. Cornelia Linde and Dorian Greenbaum (trans.)** *The Nativities of Mohammed and Martin Luther, 1604*; **J. Cornelia Linde and Dorian Greenbaum (trans.)** *The Nativity of Augustus*; **John Meeks** *Introduction: Kepler and the Art of Weather Prognostication*; **John Meeks (trans.)** *Kepler's Weather Calendar of 1618*; **John Meeks (trans.)** *Excerpts from Kepler's Weather Calendar of 1619*; **Patrick J. Boner (trans.)** *Astrology on Trial: Kepler, Pico and the Preservation of the Aspects De stella nova: Chapters 7-9*; **J. Cornelia Linde and Dorian Greenbaum (trans.)** *On Directions*; **J. Cornelia Linde and Dorian Greenbaum (trans.)** *David Fabricius and Kepler on Astrological Theory and Doctrine, 1602*; **J. Cornelia Linde and Dorian Greenbaum (trans.)** *David Fabricius and Kepler on Fabricius's Directions, 1603-1604*; **J. Cornelia Linde and Dorian Greenbaum (trans.)** *On Aspects, 1602*; **Appendix** *A Selection of Kepler's Handwritten Charts*

Contents Vol. 15 no 1 (Spring/Summer 2011)
Miguel Querejeta *On the Eclipse of Thales, Cycles and Probabilities*; **Nicholas Campion** *The Shock of the New: The Age of Aquarius*; **Alejandro Gangui** *The Barolo Palace: Medieval Astronomy in the Streets of Buenos Aires*; **Nicholas Campion and John Frawley** *Research Note: A Horoscope by André Breton*

Contents Vol. 15 no 2 (Autumn/Winter 2011)
Liz Greene *Heavenly Hosts: Angelic Intermediaries as Soul-Gates*; **Pamela Armstrong** *Ritual Ornamentation—From the Secular to the Religious*; **Paul Cheshire** *William Gilbert: Macrocosmal Astrologer in an Age of Revolution*; **Sylwia Konarska-Zimnicka** *Astrologia Licita? Astrologia Illicita? The Perception of Astrology at Kraków University in the Fifteenth Century*; **John Frawley** *Research Note: William Blake and Antares*

Contents Volume 16 No 1/2 (Spring/Summer Autumn/Winter 2012)
Nicholas Campion, *Editorial: The Inspiration of Astronomical Phenomena*; **Chris Impey**, *The Inspiration of Astronomical Phenomena*; **Ulisses Barres de Almeida**, *What are these sparks of infinite clarity? And what am I? So I pry*; BATH AND THE HERSCHELS: **Michael Hoskin**, *William Herschel's Wonderful Decade, 1781–1790*; **Francis Ring**, *The Bath Philosophical Society and its influence on William Herschel's career*; **Roberta J.M. Olson and Jay M. Pasachoff**, *The Comets of Caroline Herschel, Sleuth of the Skies at Slough*; HISTORY AND CULTURE: **V.F. Polcaro and A. Martocchia**, *Guidelines for a social history*

of Astronomy; **Euan MacKie**, *A new look at the astronomy and geometry of Stonehenge*; **Leonid Marsadolov**, *Archaeoastronomical Aspects of the Archaeological Monuments of Siberia*; **Christian Etheridge**, *A systematic re-evaluation of the sources of Old Norse astronomy*; **Aidan Foster**, *Hierophanies in the Vinland Sagas: Images of a New World*; **Inga Elmqvist Söderlund**, *Inspiration from antique heroic deeds: Hercules as an astronomer*; **Patricia Aakhus**, *Astral Magic and Adelard of Bath's Liber Prestigiorum; or Why Werewolves Change at the Full Moon*; **David Pankenier**, Astrology for an Empire: The 'Treatise on the Celestial Offices' (ca. 100 BCE); **Steven Renshaw**, *The Inspiration of Subaru as a Symbol of Values and Traditions in Japan*;b **Daniel Armstrong**, *Citing The Saucers: Astronomy, UFOs and a persistence of vision*; **Alberto Cappi**, *The concept of gravity before Newton*; **Paul Murdin**, *Artilleryman to head of state—how astronomy inspired Francois Arago*; **Paolo Molaro and Alberto Cappi**, *Edgar Allan Poe's cosmology in* Eureka; **Voula Saridakis**, *For 'the present and future happiness of my dear Pupils'": The Astronomical and Educational Legacy of Margaret Bryan*; **Michael Rowan-Robinson**, *The invisible universe*; THE ARTS: **Arnold Wolfendale**, *The Inter-Relation of the Visual Arts and Science in General and Astronomy in Particular*; **Lynda Harris**, *Changing Images of the Milky Way during the Greco-Roman and Medieval Periods*; **Lucia Ayala**, *The Universe in images: Iconography of the Plurality of Worlds*; **Tayra M. Carmen Lanuza-Navarro**, *Astrological culture before its public: the representation of astrology in Golden Age Spanish Theatre*; **Emily Urban**, *Depicting the Heavens: The Use of Astrology in the Frescoes of Rome*; **Michael Mendillo**, *The Artistic Portrayal of the Medicean Moons in Early Astronomical Charts, Books and Paintings*; **Rolf Sinclair**, *Howard Russell Butler: Painter Extraordinary of Solar Eclipses*; **Beatriz Garcia, Estela Reynoso, Silvina Pérez Alvarez and Rubén Gabellone**, *Inspiration of Astronomy in the movies: a history of a close encounter*; **Gary Wells**, *The Moon in the Landscape: Interpreting a Theme of 19th Century Art*; **Clive Davenhall**, *The Space Art of Scriven Bolton*; **Matthew Whitehouse**, *Astronomical Organ Music*; **Aaron Plasek**, *Between Scientists, Writers and Artists: Theorising and Critiquing Knowledge-Production at the Interstices between Disciplines*; ARTISTS: **Merja Markkula**, *The Way I See the Stars: fibre art inspired by astrobiology*; **Govinda Sah**, *Beyond the Notion*; **Gisela Weimann**, *Above all the stars*; **Courtney Wrenn**, *Nebulae (emission / absorption)*; **Toby MacLennan**, *Presentation of Playing the Stars*; **Felicity Spear**, *Extending vision: sky-situated knowledge and the artist's eye*; **Vanessa Stanley**, *Surveillance-Surveillance-Surveillance*; **Jim Cogswell**, *Molecular Delirium*.

Contents Vol. 17 no 1 (Spring/Summer 2013)
Clifford J. Cunningham and Günther Oestmann *Classical Deities in Astronomy: The Employment of Verse to Commemorate the Discovery of the Planets Uranus, Ceres, Pallas, Juno, and Vesta*; **Dorian Knight** *A Reinvestigation Into Astronomical Motifs in Eddic Poetry*; **Karen Smyth** *'I specially note their Astronomie, philosophie, and other parts of profound or cunning art': The Use of Cosmos Registers by Chaucer and Others*; **Kirk Little** *Spellbound: The Astrological Imagination of Washington Irving*; **Guiliano Masola and Nicola Reggiani** *Σελήνη Τοξότη: Business and Astrology in the Papyri*; **Reinhard Mussik** *Research Note: Weltall, Erde, Mensch and Marxist Cosmology in East Germany*

www.ingramcontent.com/pod-product-compliance
Lightning Source LLC
Chambersburg PA
CBHW071458080526
44587CB00014B/2148